Discrete-Event Simulation

Series Editor
Jean-Charles Pomerol

Discrete-Event Simulation

Concepts and Production in Arena

Abdessalem Jerbi

iSTE

WILEY

First published 2024 in Great Britain and the United States by ISTE Ltd and John Wiley & Sons, Inc.

ISTE Ltd
27-37 St George's Road
London SW19 4EU
UK

www.iste.co.uk

John Wiley & Sons, Inc.
111 River Street
Hoboken, NJ 07030
USA

www.wiley.com

Library of Congress Control Number: 2024943359

British Library Cataloguing-in-Publication Data
A CIP record for this book is available from the British Library
ISBN 978-1-78630-974-7

Contents

Preface

Simulation is a process which consists of designing a model of the (real) system studied, carrying out experiments on this model (and not calculations) and interpreting the observations provided by the progress of the model in order to formulate decisions relating to the system. The goal may be to understand the dynamic behavior of the system, to compare different configurations of this system, to evaluate different control strategies or even to evaluate and optimize the performance of the studied system.

Simulation can be used in different fields such as:

– production flow systems for:
  - balancing assembly lines,
  - the design of transfer systems between production stations,
  - the sizing of workshop stocks,
  - the comparison of queue control policies;
– logistics flows and transport systems for:
  - warehouse design and sizing,
  - the sizing of a truck fleet,
  - the study of procedures for controlling the flow of vehicles in circulation;
– service systems such as:
  - the study of banking transactions,
  - restaurant management,
  - comparison of maintenance policies.

This work attempts to explain in detail the different stages of the discrete event simulation technique. Thanks to simple educational examples and problems inspired by practice, the reader will then be able to apply these theoretical notions of modeling and simulation using the Arena simulation software. This book also presents various discrete event simulation projects in several fields such as the manufacturing field, the public health field and the supply chain field.

August 2024

1

Simulation

1.1. Introduction

Simulation aims to imitate the behavior of real systems on a computer equipped with appropriate software. It is generally used to analyze systems and make operational or resource policy decisions. Nowadays, simulation is a tool that is becoming increasingly popular, as computers and software become increasingly efficient (Chung 2003; Fleury et al. 2006; Altiok and Melamed 2007; Kelton et al. 2015; Polenghi et al. 2018).

The study of a system leads in most cases to modeling its functioning through the establishment of mathematical or logical relationships. It is possible to use mathematical methods if these relationships are simple enough. This solution is then called an analytical solution. However, systems in reality are most often too complex to be able to apply such an evaluation. We then resort to simulation in order to estimate the desired characteristics of the model.

Let us take the example of an industrial company which seeks to expand without being sure of the potential gain generated by this expansion. Indeed, it would not be profitable for the company to invest money for an extension, then later remove this extension, if the latter is deemed unprofitable. A simulation study could enlighten decision-makers on the consequences of this extension by simulating the operation of the factory before and after extension. We then achieve results without making physical changes in the factory.

Simulation is a process which therefore consists of designing a model of a real system, carrying out experiments on this model, interpreting the observations provided by the execution of the simulation of the model and formulating decisions relating to the system. The goal of this process can be to compare different configurations of the system studied or to evaluate different strategies for its control in order to optimize its performances.

The fields of application of simulation are numerous and varied. A non-exhaustive list of problems for which simulation has proven to be a useful and powerful tool includes:

– production flow systems (Addi Ait 2000):

- machining operations: simulations of machining operations may include processes involving manually or computer-controlled factory equipment for machining, turning, bending, cutting, and welding,

- assembly operations: assembly operations simulation can cover any type of assembly line or manufacturing operation that requires the assembly of multiple components into a single part,

– logistics flows and transport systems:

- material handling equipment: material handling simulations include analysis of cranes, forklifts and automated guided vehicles,

- warehousing: warehousing simulations may involve manual or automated storage and retrieval of raw materials or finished products;

– the production of services:

- hospitals and medical clinics: models of hospitals and medical clinics can be simulated to determine the number of rooms, nurses and doctors for a particular location,

- retail stores: retail stores may need to know how many checkouts to use,

- food or entertainment facilities: entertainment facilities, such as multi-theater movie theater complexes, may be interested in the number of ticket sellers, ticket checkers or concession stand attendants to employ,

- information technology: information technology models generally concern the number and type of networks or support resources to be made available,

- customer ordering systems: customer ordering systems may need to know how many customer order representatives are needed on duty.

1.2. Advantages of simulation

1.2.1. *Compressed time experimentation*

As the model is simulated on a computer, experimental simulations can be carried out in compressed time. This is a major advantage because some processes can take months or even years to complete. Long system processing times can make robust analysis difficult to achieve. With a computer model, the operation of long processes can be simulated in seconds. Additionally, multiple repetitions of each simulation can easily be performed to increase the statistical reliability of the analysis (Kelton et al. 2015).

1.2.2. *Reduced analytical requirements*

Before the existence of digital simulation, we were forced to use other tools that were more analytically demanding. Even then, only simple systems involving probabilistic elements could be analyzed. More complex systems were strictly the domain of the mathematical researcher or operations research analyst. Furthermore, systems could only be analyzed with a static approach at a given point in time. On the other hand, the advent of simulation methodologies has made it possible to study systems dynamically in real time. Additionally, the development of simulation-specific software has eliminated many of the complicated basic calculations and programming requirements that might otherwise have been necessary. These reduced analytical requirements have made it possible to analyze many more different types of systems than before with a greater variety of experiments (Kelton et al. 2015; Polenghi et al. 2018).

1.2.3. *Easy to demonstrate models*

Most simulation software has the ability to dynamically animate the operation of the model. Animation is useful both for debugging the model and for verifying and demonstrating its operation. Animation-based debugging makes it easy to observe flaws in model logic. Using animation during a presentation can help establish the credibility of the model through the dynamic demonstration of how the system model handles different situations. Without the capability of animation, we would be limited to less effective textual and digital presentations (Kelton et al. 2015).

1.3. Disadvantages of simulation

Although simulation has many advantages, there are also some disadvantages. These drawbacks are not really associated directly with the modeling and analysis of a system, but rather with the expectations associated with simulation projects (Chung 2003; Fleury et al. 2006; Melamed and Rutgers 2007). These disadvantages are as follows:

– Under no circumstances can simulation give accurate results when the input data are inaccurate: no matter how good a model is, if its input data are not accurate, we cannot reasonably expect to obtain accurate output data. Unfortunately, the data collection phase is considered the most difficult part of the simulation process.

Nevertheless, it is common that little time is allocated to this phase. In many cases of simulations, historical data of questionable quality have been accepted in order to save input data collection time. Too often, the exact nature or conditions under which these data are collected are unknown.

– Simulation alone cannot solve problems: some managers may believe that carrying out a model simulation project is enough to solve the problem they are facing. However, simulation by itself cannot actually solve the problem. It can provide direction for potential solutions to resolve the problem and it is up to the manager to implement the proposed changes.

1.4. Concepts relating to simulation

1.4.1. *System concept*

Different definitions have been given to the word "system" in the field of simulation (Kelton et al. 2015). Here are some examples:

– a system is a set of components linked together;

– a system is an organized set of functional elements;

– a system is a combination of parts that coordinate to achieve a result so as to form a set;

– a system is a combination of people, machines, materials and information intended to satisfy a given objective.

All these definitions have in common a few key words which characterize the notion of system. Indeed, a system is characterized by ordered parts which compose it. Each of these parts has its own laws and a certain independence. These parts also have links or relationships between them. Furthermore, the whole thing, or the system, changes over time and is influenced by the environment in which it exists and which reacts on it. Finally, this set is most often subject to constraints and only exists to achieve a goal.

In summary, a system can be defined by the knowledge of its parts or its components, the laws specific to each component and the interactions which determine its purpose.

Example: in a production plant, the parts that make up the system could be the workforce and departments of purchasing and supply, inventory management, manufacturing and production scheduling, sales and administrative. Each of these services has its own operating laws and is partially independent of the others. But also, they interact with each other. Knowledge of these interactions and their application to the system will significantly influence the goal to be achieved by the system such as profitability, investment, profit, etc.

1.4.2. *Model and modeling*

A model is a representation of a system whose goal is to explain and/or predict certain aspects of its behavior. This representation is more or less faithful, because, on the one hand, the model must be sufficiently complete in order to be able to answer the various questions that can be asked about the system it represents and, on the other hand, it must not be too complex to be easily manipulated (Kelton et al. 2015).

When modeling a system, any modeler must ensure that the boundaries of their model are clearly defined through the internal variables, inputs and necessary outputs. In addition, they must judge the level of detail to incorporate into the model.

The two methods of progressive refinement and submodel accumulation are commonly used for modeling. Progressive refinement proceeds through iterations towards increasingly increasing levels of complexity. This evolutionary process makes it possible to take into account the inadequacies of the model at a given stage and to improve it at the next stage. The accumulation of submodels is used when the system to be studied is large. In this case, the system is divided into distinct subsystems in order to control the complexity of the system. All problems can manifest themselves in the integration stage, which leads to often delicate interface problems.

There are several classifications of simulation models. However, the most used consists of classifying them according to three dimensions, which are as follows:

– Static or dynamic: static models have no interaction with time, which plays an essential role in dynamic models.

– Continuous or discrete: the state of the continuous system can change continuously over time, such as the level of a reservoir as water enters and exits. In contrast, state changes in a discrete model can only occur at specific times. This is the case of a manufacturing system where parts arrive and leave at specific times and where machines stop and start again at specific times. It is possible that the same model has both continuous and discrete elements of change. This model is then called a mixed continuous-discrete model; for example, the case of a system for extruding PVC evacuation tubes and cutting them into units of four meters in length.

– Deterministic or stochastic: models with constant inputs and variables are considered deterministic. On the other hand, stochastic models work with at least some random inputs.

– Terminating or non-terminating: terminating and non-terminating systems are distinguished by:

- initial starting conditions: terminating systems generally begin each time period without any influence from the previous period. Many systems that use client-type inputs are considered terminating-type systems. A bank, for example, does not let its customers stay overnight on its premises. This means that every new day or period of time, the system starts empty. The non-terminating system, on the other hand, can start with entries already present in the system from the previous period;

- the existence of a natural termination event: this may be the shutdown of the system at a specific time or the end of a period of activity which is of particular interest. Stores and restaurants that typically close at the end of the day are terminating-type systems. In fact, the closure constitutes a termination event and all customers are forced to leave the store. So, in each new period, the system starts empty. Airline ticket counters can also be modeled as terminating systems. In this case, the period of interest may be the time between the arrival of the plane and its departure. Thus, the final event could be the departure of the plane. The non-existence of natural stopping events means that the system can be classified as non-terminating. Many manufacturing systems operate continuously 24/7. The only time they stop is for annual maintenance. Other manufacturing systems shut down after one or two shifts per day. However, these systems keep work in progress, waiting between shutdowns. Hospitals do not normally close, although activity may be reduced overnight.

1.5. Components of a simulation model

1.5.1. *Entities*

The first type of component of a simulation model corresponds to entities. These entities are the dynamic objects of the simulation: they move for a while, and then are eliminated when they leave the system. As they move, they change state, affect and are affected by other entities as well as the state of the system, and ultimately affect system performance metrics. In many cases, particularly those involving service systems, the entity may be a customer. In other cases, entities may be objects. In a factory, entities are components waiting to be processed (Altiok and Melamed 2007; Kelton et al. 2015).

1.5.1.1. *Entity batches*

The number of entities arriving in the system at the same time is called the batch size. In some systems, the batch size is always one. In others, entities may arrive in groups of different sizes. For example, families who go to the cinema arrive in batches. Batch sizes can be two, three, four or more.

1.5.1.2. *Entity inter-arrival times*

The time between batch arrivals is called the inter-arrival time. It does not matter whether the batch size is one or more. The previous batch may have included only one entity, while the next batch may have several. The inter-arrival time is also the reciprocal of the arrival rate. When collecting the arrival of entities, it is generally easier to collect the inter-arrival time of batches. However, some historical data may be in the form of arrival rates. The inter-arrival time is considered as input data that must be provided to the model (Addi Ait 2000).

1.5.2. *Attributes*

Attributes are common characteristics that are assigned to entities. However, they may have specific values which may differ from one entity to another. It is up to the modeler to determine which attributes entities need, name them, assign values to them, modify them if necessary and then use them when appropriate (Altiok and Melamed 2007; Kelton et al. 2015).

Even if the entity attribute has the same name, it can have as many different values as there are entities. Let us take the example of the "ArriveTime" attribute which relates to the arrival time of the entity. This attribute stores the arrival time of the entity in the simulation system. Thus, each entity will have a unique value in its "ArriveTime" attribute. Some entity attributes may have the same value. Indeed, in the case of airline passengers, the "PassengerType" attribute could contain a value corresponding to the type of passenger. So, a value of 1 in "PassengerType" could represent a first class passenger and a value of 2 could represent an economy class passenger. In the simulation model, this attribute would be used to prioritize passenger boarding. So, for example, 30% of the entities in a simulation could have a value of 1 for the "PassengerType" attribute, and the remaining 70% would have a value of 2.

1.5.3. *Variables*

Unlike the attribute, a variable does not relate to an entity but rather to a characteristic of the system. You can have many variables in a simulation model, but each one must be unique. Two types of variables can coexist in a simulation model. The first type concerns predefined variables built into the simulation software such as the number of entities in the queue, the number of busy servers, the current time of the simulation clock, etc. The second type relates to user-defined variables such as average service time, travel time, etc. These variables are accessible to all entities, and many of them can be modified by any entity.

Variables are used for a wide variety of purposes. For example, the travel time between two stations in a model can be the same throughout the model; while a variable called "TransferTime" can be defined and set to the appropriate value, then used wherever it is necessary. Variables can also represent something that changes value during simulation, such as the number of parts in a certain area of the model. This variable will be incremented by one part when it enters this zone and decremented when a part leaves it (Altiok and Melamed 2007; Kelton et al. 2015).

1.5.4. *Resources*

Resources process or serve entities within the system. Here are some examples of resources:

– customer service representatives;

– hairdressers;

– counter agents;

– machines;

– ATMs.

In simple models, resources can have two states. They are then either free or busy. Resources are free when they are available for processing, but there are no waiting entities. Resources are busy when they are processing entities. In more complex models, resources may also be temporarily idle or down. Inactive resources are unavailable for scheduled work breaks, meals, preventive maintenance periods, etc. Down or failing resources correspond to broken machines or even inoperable equipment.

Resources take some processing time to serve entities, for example, the time to prepare an order and receive payment or to machine a part. Processing time is also frequently referred to as the processing time or service time. This time is considered input data that would normally be collected by observation or otherwise acquired (Altiok and Melamed 2007; Kelton et al. 2015).

1.5.5. *Waiting queues*

Entities typically wait in a queue until it is their turn to be processed. These queues are managed by priority rules. Queue priority means that the order of entities in the queue can change depending on the priority scheme (Altiok and Melamed 2007; Kelton et al. 2015). There are many queue priority rules, including:

- first in first out (FIFO);
- last in first out (LIFO);
- shortest processing time (SPT);
- longest processing time (LPT);
- lowest value first (LVF);
- highest value first (HVF).

1.5.5.1. *FIFO, LIFO*

In most simple systems, entities are served based on a first-come, first-served rule (FIFO). In this rule, the first entity in the queue will be served before other arriving entities. This type of priority rule is most often encountered in service-type systems.

LIFO, another type of priority rule, is the direct opposite of FIFO. This means that whoever entered the queue last will be the first to be processed. The use of LIFO is nowhere near as common as that of FIFO. Most LIFO applications impose some sort of penalty where the least senior members of the queue face unwanted treatment.

1.5.5.2. *SPT, LPT*

Two other queue priority rules that we may encounter are SPT and LPT. The SPT priority rule can be used effectively when there is a service limit for the system. In this case, customers who have the shortest processing time are sent to the front of the queue and are processed first. The LPT priority rule means that customers with the most complex transactions are processed first. This priority rule is much less common in service systems. It is only encountered when the longer processing time presents a much greater economic benefit to the system.

1.5.5.3. *LVF, HVF*

LVF priorities are often used to model commercial passenger transportation systems. Passengers can be classified into first, second and third classes. The class value can then be 1, 2, and 3. The priority in the LVF queue makes the first class at the front of the queue, the second class in the middle and the third class at the end. Second-class passengers can only receive service if there are no first class customers in the queue. Similarly, third-class passengers can only be served if there are no first- or second-class passengers in the queue. Every time a first-class passenger arrives in the system, they are automatically placed at the front of the queue. Similarly, any second-class passenger will automatically enter the queue behind any first-class passenger, but before any third-class passenger.

The final queue priority method is the HVF priority scheme. Here, each customer is assigned a value which rearranges the position of customers in the queue in descending order of that value. This type of priority in a queue can be used when some people are regular customers. In this situation, the service system may want to prioritize these customers.

It is possible that none of the queue priority or ranking criteria rules adequately model the type of queue priority that the actual system uses. In the case where the real system uses a complicated queue priority rule, most simulation languages offer the ability to program all the necessary calculations. These rules are generally called user-defined rules.

1.5.6. *Statistical accumulators and performance measures*

Users of simulation software are primarily interested in the performance of the system model. To ensure this, they must calculate some sort of output measure to possibly compare it to other alternative forms of the model (Altiok and Melamed 2007; Kelton et al. 2015).

To collect output performance measures during simulation, we must keep track of various intermediate variables in statistical accumulators as the simulation progresses. All these accumulators should be initialized to 0. When any event happens during the simulation, the concerned accumulator should be updated appropriately. Simulation software supports most statistical accumulators that may be collected. In some situations, it is up to modelers to define their own accumulators.

Several performance measures are used in a simulation. The most commonly used of these are:

– the total time in the system;

– the waiting time;

– the number of entities in the queue;

– the resource utilization.

1.5.6.1. *Total time in the system*

This involves measuring the time the entity spends in the system. This time begins when the entity arrives in the system and ends when it leaves the system. It is more common to use the mean flow time (MFT) for all entities. The mathematical representation of this average time is as follows:

$$\text{MFT} = \frac{\sum_{i=1}^{n} T_i}{n} \quad [1.1]$$

where T_i is the system time for an individual entity (arrival time – departure time) and n is the number of entities that are processed by the system.

1.5.6.2. *Waiting time*

Waiting time is also an observational measure. It is similar to overhead, but it only takes into account the time the entity spends in the queue. Some people do not favor waiting time, because it is the most unacceptable time. Indeed, many customers are partially unsatisfied when the service time begins, even though the service time itself may be long. The formula for mean waiting time (MWT) is:

$$\text{MWT} = \frac{\sum_{i=1}^{n} W_i}{n} \quad [1.2]$$

where W_i is the waiting time for an individual entity (queue arrival time – service start time) and n is the number of entities that are processed in the queue.

1.5.6.3. *Number of entities in queue*

The number of entities in the queue is a time-dependent statistic. The average number in queue is the average number of entities that we would expect to see in the queue at any time during the considered period. At any time, the queue actually contains a discrete number of entities. However, since the average number in the queue is an average value, it will usually result in a number that has a fractional value. For lightly loaded queues, it is actually possible for the average number in the queue to be less than 1. The formula for calculating the average number in queue (ANQ) is the following:

$$\text{ANQ} = \frac{\int_0^T Q\,dt}{T} \quad [1.3]$$

where Q is the number of entities in the queue for a given duration, dt is the duration that Q is observed and T is the total duration of the simulation .

1.5.6.4. *Resource utilization*

Resource utilization is also a time-dependent statistic. At any time, a resource can be either idle or busy. The inactive state corresponds to a resource utilization level of 0 and, naturally, the busy state corresponds to a resource utilization level of 1. The duration during which the resource is either at level 0 either at level 1 is a function of the entities that enter the system. The formula for the average resource utilization (Util) is

$$\text{Util} = \frac{\int_0^T \text{Bdt}}{\text{T}} \qquad [1.4]$$

where B is either 0 for idle or 1 for busy, dt is the duration that B is observed and T is the total duration of the simulation.

1.5.7. *Event, activity and process*

The simulated system passes from one state to another during the simulation. The entities then undergo operations and move through the system. Resources, for their part, execute operations and change state. Basically, the entire simulation process is event-driven. An event is something that occurs at an instant in the simulated time and can modify attributes, variables or statistical accumulators (Kelton et al. 2015).

In a service system, customers may queue for different time durations or have different services or quantities of goods to purchase. Therefore, the model must at least include the following events:

– customer arrivals in the processing area;

– processing customer requests;

– payments for goods.

At each event, the objects, entities and resources concerned engage in operations. These operations which are initiated or terminated by events are called activities. Any activity is therefore limited at its beginning and its end by an event.

The grouping of events according to the chronological order in which they occur is a process. This also includes activities. An activity is then triggered by a given event and stopped by another event. A process is therefore a sequence of events ordered in time and which delimit the durations of several activities (Figure 1.1).

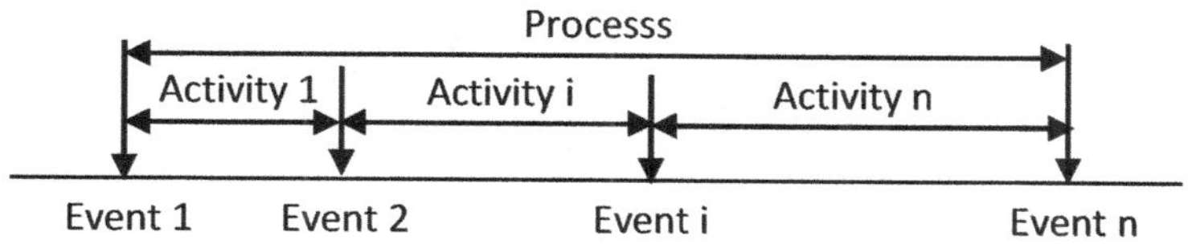

Figure 1.1. *Relationship between event, activity and process*

1.6. Basic principles of system representation

An organizational chart is an essential tool for defining the system. It then helps us to obtain a fundamental understanding of the logic of the system and graphically describes its main components and events. It also illustrates how input and output data play a role in the model.

1.6.1. *Standard flowchart symbols*

There are four basic symbols for flowchart processes which are as follows:

– oval;

– rectangle;

– inclined parallelogram;

– diamond.

The first guideline for using these symbols is that they should be arranged such that the sequence of processes flows down and to the right as much as possible. The second guideline is that a given symbol should have only one connection path to the other symbols. The only exception to this rule is the decision icon, which has two different exit paths. However, even with the decision icon, only one exit route can be taken at any given time. Finally, it is useful to try to keep the organization chart on as few sheets as possible.

1.6.1.1. *Start and stop oval*

The oval symbol is used to designate start and stop processes. The start symbol is the first symbol that must be present on the flowchart. The start symbol usually has only one output connector. The end symbol is normally the last symbol present on the flowchart. Even if the process has several possible paths, there should only be one stop or end symbol (Figure 1.2).

Figure 1.2. *Start and end symbols in a flowchart*

1.6.1.2. *Process rectangle*

The rectangle is used to designate general-purpose processes that are not specifically covered by one of the other flowchart symbols. The process rectangle is normally entered from the top or the left. Its exit can occur from the bottom or from the right side (Figure 1.3).

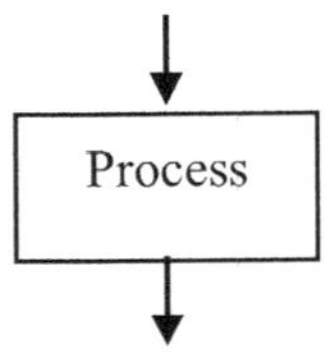

Figure 1.3. *Process symbol in a flowchart*

1.6.1.3. *Inclined input/output parallelogram*

The inclined parallelogram is used for processes that involve some sort of input or output. Entry into the inclined parallelogram is normally from the top or left side. Exit is from the bottom or right side. An example of an input process symbol in the flow diagram of a simulation program would be the creation or arrival of entities in the system (Figure 1.4).

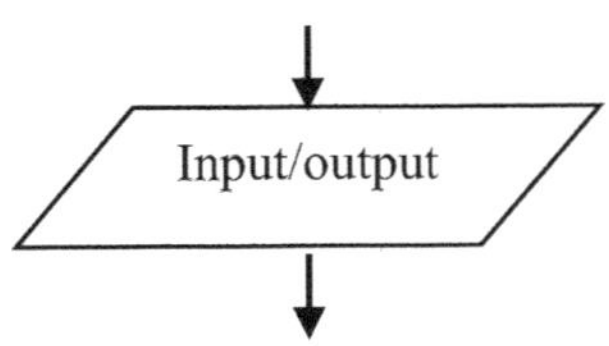

Figure 1.4. *Input/output symbol in a flowchart*

1.6.1.4. *Decision diamond*

The diamond is used to represent a decision in the logic of the organization. The diamond has one input connector but two output connectors. The single input connector should land on the top of the diamond symbol. The output connectors can exit through one of the side vertices or through the bottom vertex. These output connectors should be specifically labeled as "true" or "false" or "yes" or "no". As these responses are mutually exclusive, only one exit path can be taken at any given time (Figure 1.5).

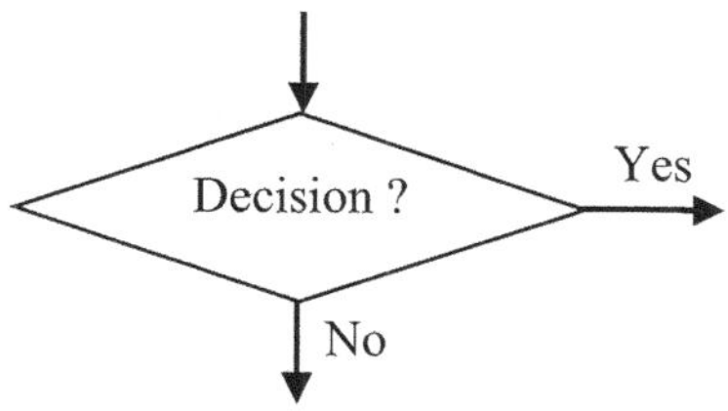

Figure 1.5. *Decision symbol in a flowchart*

1.6.2. *Flowchart example*

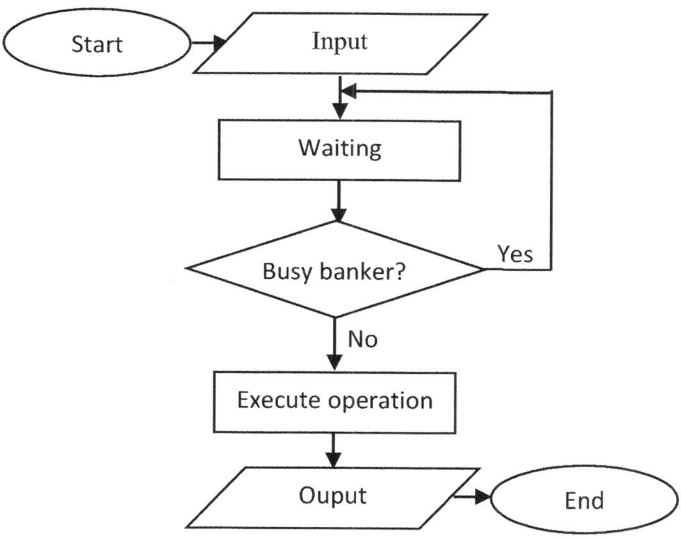

Figure 1.6. *Flowchart example*

Figure 1.6 is a single-queue, single-server system that represents the operation of a bank teller.

The system flow diagram begins with the starting oval. The following symbol is an inclined input/output parallelogram, which represents the customer entering the system. Once customers enter the system, they are made to wait in the queue. Waiting in this queue is a process represented by a rectangle. The customer then determines whether the banker is busy or not. This operation is represented by a diamond. If the banker is free, the customer carries out the process of the operation represented by another rectangle. On the other hand, if the banker is busy, the customer remains in the queue. When the customer finally completes the operation, they exit the system. This operation is represented by another inclined parallelogram. The flow diagram then ends with an oval end symbol.

Notice how the flow of the process is oriented to the right and down as much as possible. The only exception where the flow is not downward is when the customer remains in the queue. It should also be noted that most symbols have only one input and one output. The only exceptions are starting with a single exit, stopping with a single entry, and decision. Although the decision has two exits, only one exit can be taken at a time. Either the banker is busy or they are not. Also note that when the banker is busy, the flow line does not return directly to the queue. Instead, the flow line intersects the other flow line that goes into the queue. This preserves the principle that the queue symbol can only be entered in one place.

1.7. Manual simulation

This is the same example of the bank teller discussed previously. The following inter-arrival and service times were observed in a single-server, single-queue system:

– inter-arrival times (minutes): 1, 2, 4, 2, 8, 2, 4, 3;

– service time (minutes): 2, 7, 4, 1, 3, 2, 1, 3.

Arrival number	**Arriving time**	**Service start time**	**Service time**	**End of service time**	**Total time**
1	1	1	2	3	2
2	3	3	7	10	7
3	7	10	4	14	7
4	9	14	1	15	6
5	17	17	3	20	3
6	19	20	2	22	3
7	23	–	–	–	–

Table 1.1. *Manual simulation of the example*

Let us try to calculate summary statistics for the average wait time, average total time and average utilization for a 20-minute simulation (Table 1.1).

The first event is the arrival of the first customer which occurs at t =1 minute. Since the queue is empty and the banker is inactive, this customer can immediately seize it to start the service. This means that the service time also starts at t = 1 minute.

The start of the service is the second event in the system. The service time for the first customer is two minutes; this means that the end of service occurs at t = 3 minutes. The end of this customer's service is the third event in the system. It should be noted that the total time this customer is in the system corresponds to the difference between their service end time at t = 3 minutes minus their arrival time at t = 1 minute which corresponds to a total duration of two minutes.

The fourth event that occurs is the arrival of the second customer at t = 3 minutes. The queue being empty and the banker being inactive, this customer goes directly to the counter. The second customer's service time is scheduled for seven minutes. This means that the final service time of the entity representing this customer is programmed at t = 10 minutes.

The inter-arrival time between the second and third customers is four minutes. This third customer therefore arrives at t = 7 minutes. Thus, the next event is not the end of the second customer's service, but the arrival of the third customer. However, as the banker is busy with the second customer, the third customer is put on hold at the first position in the queue.

The inter-arrival time between the third and fourth customers is two minutes. This means that the latter arrives at t = 9 minutes and enters the queue behind the third customer. There are now two customers in the queue.

Now, two other events could happen. The second customer's service time may end, or a fifth customer may arrive in the system. It turns out that the inter-arrival time of this fifth customer is two minutes. This means that the end of service for the second customer will be the following event at t = 10 minutes. It is then possible to calculate the overhead of the second customer in the same way as for the first. As the second customer did not wait in the queue before being served, their total time is the same as their service time, which is equal to seven minutes.

At t = 10 minutes, another event occurs. Indeed, as soon as the second customer has finished, the third customer immediately seizes the banker. With this customer's service now in progress, only the fourth customer is now on hold.

The third customer has a service time of four minutes. This means that its service time will be over at t = 14 minutes. And since the inter-arrival time of the fifth customer is eight minutes, the service time of the third customer will be finished before the arrival of the fifth customer, at t = 17 minutes. The overhead calculation for the third customer is slightly different from the first two. Indeed, in addition to their service time of four minutes, this customer waited in the queue for three minutes. In this case, the total time is therefore 14 – 7 = 7 minutes.

As the fourth customer was waiting in the queue when the third customer finished, he immediately leaves the queue and seizes the banker at t = 14 minutes. The queue is now empty.

The next event turns out to be the end of the fourth customer's service time. This customer has a short service time of only one minute. However, they also waited in the queue and their total time is 15 – 9 = 6 minutes.

The banker is now inactive at t =15 minutes, and the only event that can happen after that is another arrival. The next event is then the arrival of the fifth customer at t = 17 minutes. With no one in the queue, this customer immediately seizes the banker.

Subsequently, two different events could occur. The fifth customer's service time may end, or a new customer may arrive. The inter-arrival time of the sixth customer is two minutes, while the service time of the fifth customer is three minutes. This means that the next event is the arrival of the sixth customer at t = 19 minutes. As the banker is busy with the fifth customer, the sixth customer enters the queue.

The inter-arrival time of the seventh customer is four minutes for an arrival scheduled at t = 23 minutes, while the service of the fifth customer ends at t = 20 minutes. The next event is therefore the end of service for the fifth customer and the overhead can be calculated for this customer as 20 – 17 = 3 minutes. Additionally, as the sixth customer is waiting in the queue, they can immediately seize the banker at t = 20 minutes.

The next event can be either the arrival of the seventh customer or the end of service for the sixth customer. The inter-arrival time of the seventh customer is four minutes for an arrival at t = 23 minutes. The service time for the sixth customer is two minutes. This means that the next event would be the end of service for this customer at t = 22 minutes.

In fact, in the model, it was specified to calculate the summary statistics based on a simulation time of 20 minutes. This means that the arrival of the seventh customer at t = 23 minutes as well as the end of service of the sixth customer at t = 22 minutes must not be taken into account.

1.7.1. *Calculation of average total time*

To calculate the average total time in the system, simply add up all the individual total times in the system and divide them by the number of entities processed during the simulation. This number of entities is equal to 5:

$$\text{MFT} = \frac{2+7+7+6+3}{5} = 5 \text{ minutes} \qquad [1.5]$$

For all entities that have been processed by the system, this average total time is five minutes.

1.7.2. *Calculation of average waiting time*

Customers 1 and 2 arrive respectively at t = 1 and t = 3 minutes with a banker available immediately. No waiting can therefore be considered.

When customer 3 arrives at t = 7 minutes, they must wait in the queue until customer 4 arrives at t = 9 minutes. This means that for two minutes between seven and nine minutes, there was a customer in the queue:

1 customer in line × (9 – 7) minutes = 2 customers.minute

The next relevant event is the end of customer 2's service at t = 10 minutes. This means that both customers 3 and 4 waited in the queue for the period between nine and 10 minutes:

2 customers in queue × (10 – 9) minutes = 2 customers.minute

At t = 10 minutes, there is only customer 4 left in the queue. The next relevant event is the end of service of customer 3 at t = 14 minutes. This means that customer 4 waited alone in the queue between 10 and 14 minutes:

1 customer in queue × (14 – 10) minutes = 4 customers.minute

At t = 14 minutes, there is no one left in the queue. The next event is the arrival of customer 5 at t = 17 minutes. As the queue is empty and the banker is inactive, customer 5 does not have to wait. The next event that occurs is the arrival of customer 6 at t = 19 minutes. The banker being busy, this customer enters the queue. The next event is the end of customer 5's service time at t = 20 minutes, which is also the end time of the simulation. This event causes client 6 to leave the queue. This was then the only customer in the queue for one minute:

1 customer in queue × (20 – 19) minutes = 1 customer.minute

The average time in queue (AWT), can now be calculated by adding the periods that entities waited in queue and dividing by the simulation time:

$$\text{AWT} = \frac{2+2+4+1}{20} = 0.45 \text{ minute} \qquad [1.6]$$

This means that the average number of customers in the queue at any given time is almost half of a customer. Obviously, this is a mathematical representation, as it is impossible to see half of a customer. However, this indicates that the system is not very busy if there is only one "half a person" in the queue at any given time. The graphical representation of the time average of the number of customers in the queue is shown in Figure 1.7.

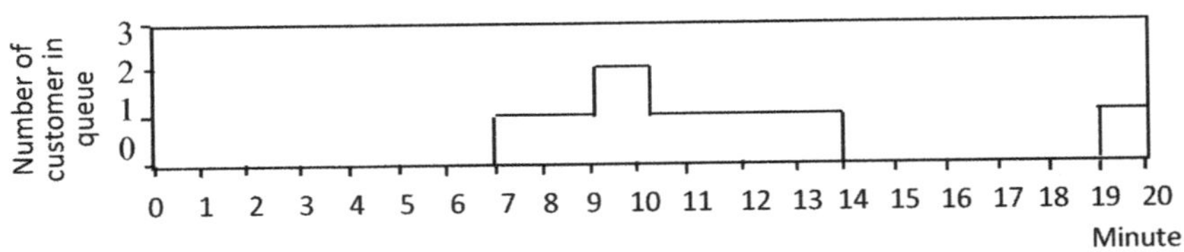

Figure 1.7. *Number of customers in queue*

1.7.3. *Calculation of average resource utilization*

Average resource utilization calculations are rather easier to perform because the single resource can only be idle or busy (Figure 1.8).

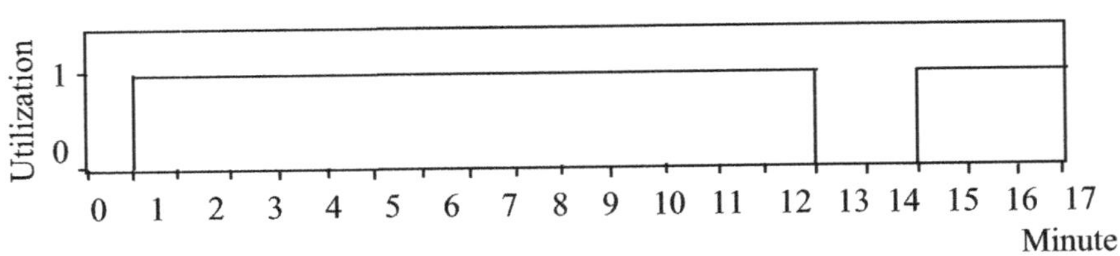

Figure 1.8. *Resource utilization*

Relevant resource utilization events start at t = 1 minute. This is the moment when customer 1 arrives and finds the queue empty and the banker idle. Thus, between zero and one minute, the utilization rate is 0:

0 resource busy × (1 – 0) minute = 0 resource.minute

The resource remains busy until customer 1's service time ends. This happens at t = 3 minutes. This area is calculated by:

1 resource busy × (3 – 1) minute = 2 resources.minute

At t = 3 minutes, customer 2 arrives and the banker is seized again until t = 10 minutes:

1 resource busy × (10– 3) minutes = 7 resources.minute

At the end of customer 2's service time, customer 3 is already waiting in the queue. Customer 3 immediately enters the resource for its four-minute service time, up to t = 14 minutes in the simulation:

1 resource busy × (14 – 10) minutes = 4 resources.minute

At the end of customer 3's service time, customer 4 waits in line. Customer 4 immediately grabs the resource for its one minute service time, up to t = 15 minutes in the simulation:

1 resource busy × (15 – 14) minutes = 1 resource.minute

At t =15 minutes, there are no more customers waiting in line. The next customer arrives at t = 17 minutes:

0 resource busy × (17–15) minutes = 0 resource.minute

At t = 17 minutes, customer 5 arrives and immediately grabs the resource and uses it for three minutes, until t = 20 minutes in the simulation:

1 resource busy × (20 – 17) minutes = 3 resources.minute

The average utilization of the single resource, "Util", can be calculated as follows:

$$\text{Util} = \frac{2+7+4+1+3}{20} = 0.85 \quad [1.7]$$

The average resource utilization rate is 0.85. This resource is therefore used for 85% of the time it is available. Such an average can be considered high, which reflects the good exploitation of the resource.

2

Simulation Project

2.1. The stages of a simulation project

As for any project, the simulation must be planned in order to achieve convincing results. The stages of a simulation project almost all follow a path common to all projects (Chung 2003; Fleury et al. 2006; Altiok and Melamed 2007; Kelton et al. 2015; Polenghi et al. 2018). These steps and their progress can be illustrated by the flowchart shown in Figure 2.1.

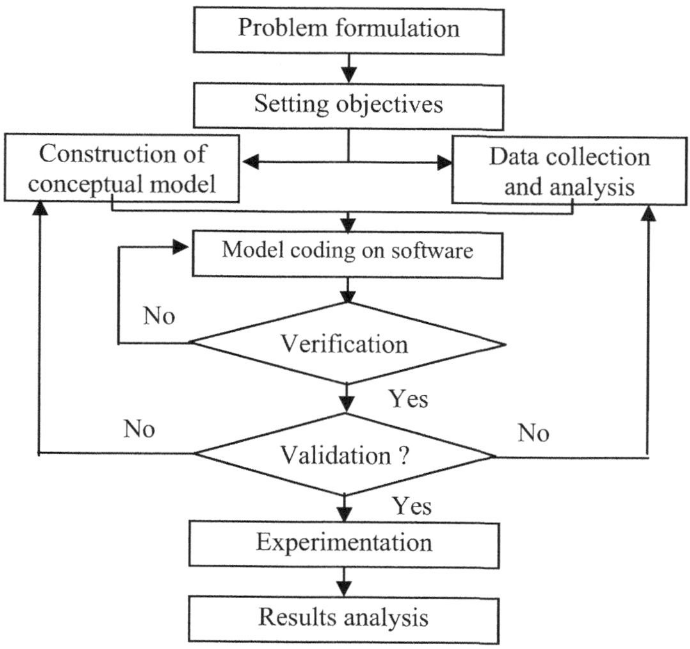

Figure 2.1. *The stages of a simulation project*

2.2. Problem formulation

The initial step in a simulation project involves ensuring that thorough attention has been paid to understanding the problem that the project aims to resolve. This activity is known as the problem formulation process. During this stage, it is suggested that a set of hypotheses be proposed, discussed and accepted by all members of the project. This problem formulation step should provide the potential audience with a clearly understandable justification for the simulation.

2.3. Setting objectives

Once the problem has been formulated, it is necessary to think about the specific objectives of the simulation project. Obviously, the objectives must be developed in close collaboration with the people commissioning the simulation study. Objectives include:

– the questions to which the simulation project should provide an answer;

– the personnel required;

– the hardware and choice of computer software;

– the various scenarios to be investigated and the expected outputs;

– the costs of the study as well as the time required.

These objectives can also be considered dynamic. During the study, we may possibly uncover additional information that could prove more profitable to pursue. Sometimes, the commissioning organization is not initially as interested as they should be.

2.4. Construction of the conceptual model

This involves formalizing the real system in the form of an abstract model (Chung 2003; Fleury et al. 2006; Polenghi et al. 2018). This model is a set of relationships between the components and structure of the system. We must start with a simple basic model, and then refine it gradually in order to obtain the model that meets the desired objectives. For example, in a first step, we start by building a first basic model including the arrivals, the queues, the arrival phenomena and the machines. In a second step, machine breakdowns and handling or transport possibilities within the system can be added. You should not think about building an excessively complex model the first time. This would increase completion times without improving the quality of the results.

2.5. Data collection and analysis

During this step, it will be necessary to inventory the data needs of the real system which are necessary for the development of the simulation model (Chung 2003; Fleury et al. 2006; Kelton et al. 2015; Polenghi et al. 2018). Ideally, one will already have the necessary data to use, which is very rare in reality. This topic is the subject of a subsequent chapter.

2.6. Software coding

The goal of this phase is to translate the conceptual model of the system into a computational model that can be used to generate experimental data. This is a two-step process. The first step of this process requires choosing the type of computer software program to use to model the system. The second step of the model translation phase consists of programming the simulation model (Chung 2003; Fleury et al. 2006; Altiok and Melamed 2007; Kelton et al. 2015; Polenghi et al. 2018).

There are currently a number of simulation software programs on the market. We can cite Arena, Automod, Simscript, Simul8, AnyLogic, etc. The number of these software programs continues to increase from year to year.

2.7. Verification

During the model coding phase, we naturally have to ensure that the simulation model has all the necessary components and that it actually works (Chung 2003; Fleury et al. 2006; Altiok and Melamed 2007; Kelton et al. 2015; Polenghi et al. 2018). In reality, it is not enough that the model works, but it must work as intended. This process is known as model verification. Another way to approach verification processes is to think of them as "you have to build the model correctly".

There is a tendency to confuse model verification with model validation. However, there is a big difference between these two stages of a simulation project. Indeed, verification is an ongoing process of ensuring that the model works as expected, while validation is the process of ensuring that the model works correctly and represents reality. Obviously, there is no point in trying to validate a simulation model if the model does not even perform as expected. Here are some recommendations for successfully completing this step:

– always consider that the model contains errors and look for them, i.e. take a destructive approach, rather than a constructive approach when creating the model;

– include people not involved in the design and implementation. Indeed, they will have enough perspective to detect trivial faults in the model;

– perform tests, such as replacing random times with constant times, testing only part of the model or testing the model under boundary conditions. To do this, we can increase the arrival rate and/or reduce the service rate to create congestion phenomena or machine starvation. We can also reduce the size of stocks to create blockages and modify the distribution of types of parts to increase the arrival of parts of less frequent types. It is also possible to increase the rate of occurrence of less frequent events, for example, breakdowns;

– use animation to analyze the model trace and check the path of the entities and the changes in state after a wait;

– avoid classic errors, particularly with regard to input data entry, arithmetic errors (parentheses, type conversion, etc.) and recording errors (arrival times of parts, counters, etc.).

2.8. Validation

Validation is the process of ensuring that a model represents reality at a given confidence level. It is in fact a question of knowing whether the model can replace the real system for the purpose of the experiment. However, for a variety of reasons, even carefully constructed models may not represent reality. This means that even well-debugged and improved models may not be suitable for conducting any type of analysis.

The ideal way to validate the simulation model is to compare its outputs with those of the real system. Several techniques can be used during this step.

2.8.1. *Sensitivity analysis*

This involves testing whether a change in the model input data results in the predictable change in its outputs. For example, if the customer arrival rate increases, the waiting time in line should also increase. It is in fact a question of checking whether the model reacts correctly to changes in the values of certain input data and if the reaction obtained is indeed what was expected.

2.8.2. *Extreme condition testing*

This involves testing whether the model behaves normally when the input data are at their extreme values. For example, if the customer arrival rate is set to its maximum value, this value should increase the waiting time in queues, the total time in the system, the number of customers in lines, etc.

2.8.3. *Validation by hypothesis testing*

This technique consists of comparing the results of the simulation model with those of the real system provided that the system already exists. A widely used method of comparison is that based on the *t*-test or Student's hypothesis test (Foucart 2000; Ycart 2002).

Example: let us take the same example discussed in Chapter 1. We seek to validate the model through the comparison of the average waiting time of a customer in front of the banker in the simulation model with the actual average waiting time in the real bank. The real average waiting time observed over five days is equal to three minutes. We then carry out five simulation replications with the simulation model. Each replication corresponds to the operation of the bank teller, for one day. These replications provide the results for the average waiting time: 1.65; 2.12; 3.51; 2.22 and 3.15 minutes. It is then a question of testing the hypotheses:

– H_0: $E[X_i] = 3$ minutes is acceptable or not;

– H_1: $E[X_i] \neq 3$ minutes,

where X_i corresponds to the random variable average waiting time during the ith replication.

The statistical method to be used is based on the Student distribution which applies to the case of small samples having a number of observations less than 30. It is then a matter of calculating the parameter "t" of the distribution and setting the level of confidence to grant to the result, 95% or 99%, and so on. Subsequently, the obtained value "t" obtained is compared to the corresponding one given in the Student table. If $t_{calculated} < t_{table}$, the hypothesis H_0 is retained and the model is validated; otherwise, it is rejected.

2.8.4. *Validation by confidence interval*

To take into account the stochastic property of the studied systems, several simulation replications must be conducted. The real or theoretical value is then compared to the confidence interval (CI) of the experimental observations obtained.

If the theoretical or real value is included in this CI, we can then confirm the validation of the developed simulation model. Otherwise, the simulation model must be reviewed.

In this case, the average of "m" observations of each replication "i" is:

$$\overline{X}_i = \frac{1}{m}\sum_{j=1}^{m} X_{ij} \qquad [2.1]$$

where X_{ij} is observation j of replication i.

The total average of the n replications is estimated as follows:

$$\overline{X} = \frac{1}{n}\sum_{i=1}^{n} X_i \qquad [2.2]$$

The CI 100(1 – α)% is obtained using Student's distribution table when n<30 or the normal distribution table when n ≥30. α being the chosen risk:

$$IC = \overline{X} \pm \eta_{\alpha/2}\frac{\sigma}{\sqrt{n}}$$

where σ is the standard deviation estimated by:

$$\sigma^2 = \frac{1}{n-1}\sum_{i}^{n}(\overline{X}_i - \overline{X}) \qquad [2.3]$$

2.9. Design of an experimental framework

This step consists of defining for each scenario studied a certain number of parameters such as:

– the duration of the simulation for each replication;

– the number of replications to be made per scenario;

– the initial state of the model.

2.10. Running the simulation and analyzing the results

Once the simulation model has been developed and simulated, it will be necessary to analyze and interpret the results obtained. The analysis of these results will aim to estimate the performance measures of the tested scenarios. Analyzing the results of a simulation generally consists of estimating the value of the mean of the outputs provided by the simulation as well as estimating its variance. These two estimates are influenced by the initial conditions with which the experiment begins (Altiok and Melamed 2007; Kelton et al. 2015). By initial state, we essentially mean the status of the queues (empty or not), that of the servers (free or busy), etc. This section is the subject of Chapter 4.

3

Data Processing

3.1. Introduction

In any simulation project, input data are essential for both model development and simulation execution. This process involves the collection of input data which can be sourced from historical records or obtained through real-time measurements.

Analyzing these data involves identifying the theoretical distribution that represents them. A common question raised by non-specialists when initiating a simulation project is why fitting data to a theoretical distribution is necessary if the data have already been collected. The basic reason is that only a sample of the actual data distribution is collected and some values are not observed; this does not mean that they do not actually exist in the system. Fitting the observed data to a theoretical distribution allows for the utilization of any value from that distribution in the simulation model. This approach is much more realistic situation than driving the simulation with only observed data values.

Collecting input data is often considered the most difficult process in conducting a simulation modeling and analysis project. This is partly because the simulation analyst may be dependent on people and operations outside the project over which there is no control.

In other situations, the required input data may not even exist. This situation is frequently encountered when the project involves the analysis of new capital equipment.

3.2. Collecting input data

When collecting the input data in real time, there are two primary methods: manual collection using timing devices or automated collection with electronic devices. With the first method, it may be useful to create a form to organize the data collection process. Alternatively, with the second method, it is possible to use a wide choice of electronic time study devices. Additionally, simple programs developed on a computer can facilitate data collection. By pressing programmed keys, arrival rates can be tracked without having to note arrival times. The advantage of this approach is that the data can be automatically saved to a file for analysis. This avoids the extremely tedious process of transcribing data from paper to computer. The use of video recording devices provides yet another opportunity for data collection when privacy issues do not need to be considered.

3.2.1. *Mode and units of time collection*

An important issue for simulation input data regarding time intervals is the choice of time unit that should be used. Typically, for inter-arrival time, absolute time or the arrival time of various entities in the system is collected and recorded. However, additional work will be required to convert the absolute time to a relative time to calculate inter-arrival times. It is therefore more effective to initially collect the data using a relative inter-arrival time approach.

Another consideration in time collection is determining the appropriate units to use. While using seconds for calculations may be precise, it can be challenging to comprehend and manage large values, such as simulating for 28,800 seconds instead of 8 hours or 480 minutes. Opting for a time unit like minutes strikes a balance, making the simulation duration more understandable. However, consistency is crucial and all data should be collected in the same unit or converted accordingly. For instance, if service times are recorded in minutes and seconds, converting seconds to fractions of minutes becomes necessary for uniformity.

3.2.2. *Deterministic and probabilistic data*

When collecting input data, it is essential to realize that there are different classifications of data. A common classification method is to determine whether the data are deterministic or probabilistic.

3.2.2.1. *Deterministic data*

Deterministic data are those where the event involving the data occurs in the same or predictable way every time. This means that this type of data only needs to be collected once, because their value never changes. Here are some examples of deterministic input processes:

– processing times of computer numerically controlled (CNC) machining programs;

– preventive maintenance intervals;

– conveyor speeds.

Processing times for CNC machining programs are deterministic because the program always takes the same amount of time to complete. These programs follow a defined number of processing steps with predetermined feed and cut speeds, with the machine tool controlled by the computer. Unless there is a machining problem, there is no possibility of deviating from the computer program.

Preventative maintenance schedules can be established with predetermined intervals. These intervals might be based on the number of components processed or a specific duration of time elapsed. In both cases, the frequency of maintenance is predetermined and can remain the same for each preventive maintenance cycle.

Likewise, most conveyors operate at a particular speed determined by the speed of the motor and the gear that drives the conveyor. This means that the conveyor speed can be set to different values, but when the conveyor is running, it runs at the set speed.

3.2.2.2. *Probabilistic data*

Unlike deterministic processes, a probabilistic process does not occur with the same regularity. In this case, the process follows a certain probabilistic distribution. Here are some examples of probabilistic data:

– inter-arrival times;

– customer service process;

– repair time.

The inter-arrival times of entities in a system are almost always probabilistic. Indeed, the duration between one entity arrival and the next can vary, making it uncertain when the subsequent entity will arrive. However, by collecting a number of inter-arrival data, it is possible to fit them in a probabilistic distribution.

Customer service times are also probabilistic. Indeed, the time required to process a customer at a service center varies depending on this customer's needs. Additionally, any time a human is involved in a service process, variations in lead times are expected. Typically, in a service process, a number of customers will be served faster than others. Likewise, processing another number of customers may take much longer time. The diagram describing the number of observations versus processing time can be used to fit a probabilistic distribution to this time.

Similar to service times, repair times are often probabilistic. Indeed, it is impossible to predict the amount of time that a problem needs to be repaired. If the problem is not serious, the repair time is expected to be short. Conversely, if the problem is serious or requires additional parts, the repair time will be long.

3.2.3. *Discrete and continuous data*

Another aspect to consider when classifying input data is whether it is discrete or continuous. Discrete data can only assume specific values, typically integers. In simulation applications, examples of discrete data include:

– the number of customers who arrive in a system as a group or batch;

– the number of jobs processed before a machine fails.

The number of customers in a batch can only take an integer value. Indeed, it is not possible to have a fraction of a customer. Similarly, the number of tasks processed by a machine before it crashes is an integer. Indeed, the work is either completed or not completed.

Continuous distributions, on the other hand, can take any value within the observed interval. This means that mixed numbers are entirely possible. Here are some examples of continuous type data:

– time between arrivals;

– service time;

– delivery time.

Obviously, the time between arrivals of entities in a system can take any value between 0 and infinity, regardless of the unit of time used. This certainly includes fractional time values. Similarly, entity service and delivery times can also take any value within a reasonable range.

3.3. Fitting a probability distribution to the variable

When we gather a sufficient observation volume of data, typically 100 or more observations, it becomes feasible to fit a theoretical probability distribution to this dataset. However, selecting one distribution over others must be justified by carrying out an adequacy test (Foucart 2000; Ycart 2002). It is worth mentioning that there is software available specifically designed to automate this statistical adjustment process (e.g. Input Analyzer in Arena).

Conversely, when the volume of data observations is not sufficient, the adjustment methods, although applicable, pose problems because of the adequacy tests which are unfortunately not adapted to these cases of reduced amounts of data. Fitting a probability distribution to the data is done in three steps which are the choice of a distribution, the estimation of the parameters of the chosen distribution and the test of adequacy of this distribution to the data.

3.3.1. *Choosing a distribution*

This initial step involves identifying whether the phenomenon is discrete or continuous. Discrete data result from counting processes, such as the number of customers arriving at a bank every hour, the number of coins received by a machine within a certain time or the number of tool changes performed on a machine daily. Continuous data result from a measurement process such as the time taken for a customer to be served, or the time taken for a part to exit the system. Furthermore, since the probability distributions are divided into discrete distributions (Poisson, binomial, geometric, etc.) and continuous distributions (uniform, exponential, normal, and triangular, etc.), this initial step can aid in selecting the appropriate type of distribution.

3.3.2. *Estimation of the parameters of the chosen distribution*

Each distribution can be described by different parameters. For example, a normal distribution is characterized by its mean and standard deviation. Thus, this task involves determining the parameters that characterize it for the selected distribution.

3.3.2.1. *The usual distributions*

Each distribution is characterized by its density function, denoted f(x). This density function checks the following points:

$$f(x) = \begin{cases} \forall x \in R, f(x) \geq 0 \\ \int_{-\infty}^{+\infty} f(x)dx = 1 \end{cases} \quad [3.1]$$

Thus, the probability that X takes a value x in an interval [a, b] is given by: P $(a \leq x \leq b) = \int_a^b f(x)dx$. It is more convenient to use the distribution function instead of its density. This distribution function is given by the following formula: $F(x) = P(X \leq x) = \int_{-\infty}^{x} f(u)du, \forall x \in R$. Thus $P(a < x \leq b) = F(b) - F(a)$ (Figure 3.1).

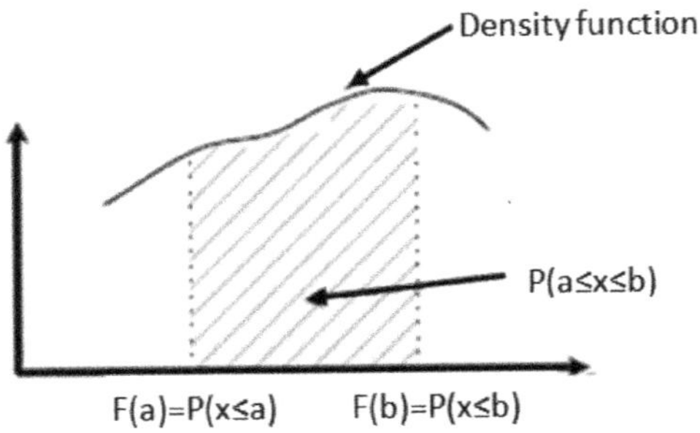

Figure 3.1. *Density and distribution function*

3.3.2.1.1. Uniform distribution

Uniform distributions are primarily used to model process input data when limited information is available about the process. All that is needed to use uniform distribution is the minimum time and maximum time taken by the process.

In a uniform distribution, every individual value has an equal probability of being observed out of all possible values (Figure 3.2). For U[a, b] or UNIF[a, b], this distribution represents equiprobable events. Its density function is as follows:

$$f(x) = \begin{cases} \frac{1}{b-a} & \text{if } x \in [a,b] \\ 0 & \text{else} \end{cases} \quad [3.2]$$

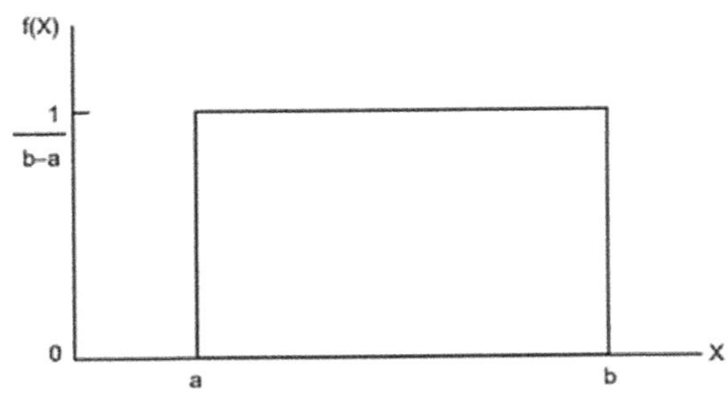

Figure 3.2. *Density function of the uniform distribution*

The formulas for the mean and variance of a uniform distribution are as follows:

$$\text{mean} = \frac{(a+b)}{2} \quad [3.3]$$

$$\text{variance} = \frac{(b-a)^2}{12} \qquad [3.4]$$

3.3.2.1.2. Normal distribution

The duration of numerous service processes conforms to a normal distribution. This phenomenon arises because many processes comprise several subprocesses. Regardless of the probability distribution of each individual subprocess, when the durations of the subprocesses are summed, the resulting durations are often normally distributed.

The normal distribution is characterized by two parameters, which are the mean and the standard deviation. The normal distribution is also symmetric, indicating an equal number of observations below and above the data mean. The pattern or distribution of observations on each side is also similar (Figure 3.3).

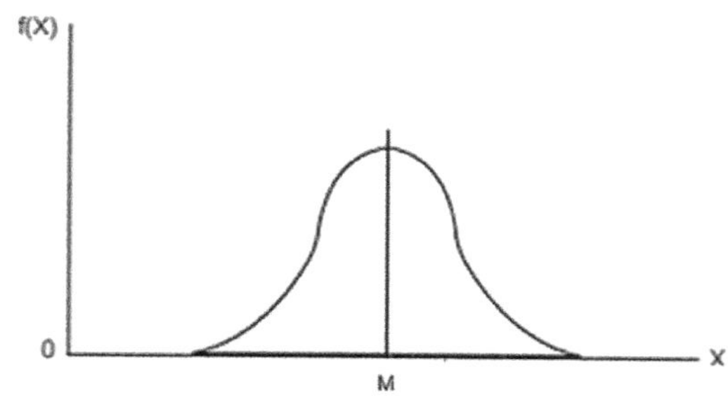

Figure 3.3. *Normal distribution density function*

For N(m, σ), its density function is as follows:

$$f(x) = \frac{1}{\sigma\sqrt{2\pi}} e^{-(x-M)^2/2\sigma^2} \qquad [3.5]$$

where M is the mean and σ is the standard deviation.

3.3.2.1.3. Triangular distribution

The triangular distribution is applicable in situations where complete knowledge of the system is lacking, but there is suspicion that the data are not uniformly distributed. The triangular distribution has three parameters, which are the minimum possible value, the most common value and the maximum possible value (Figure 3.4). Since the most common value does not necessarily have to lie equally between the minimum value and the maximum value, the triangular distribution does not have to be symmetrical. For T(a,m,b), the density function of this distribution is as follows:

$$f(x) = \begin{cases} \frac{2(x-a)}{(m-b)(b-a)} & \text{if } a \leq x \leq m \\ \frac{2(b-x)}{(b-m)(b-a)} & \text{if } m \leq x \leq b \\ 0 & \text{else} \end{cases} \quad [3.6]$$

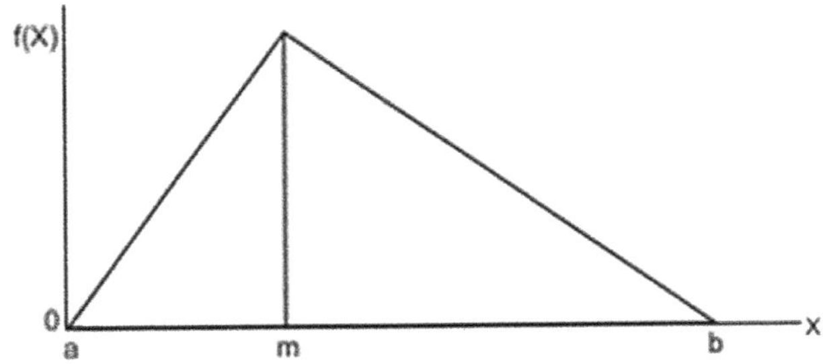

Figure 3.4. *Density function of the triangular distribution*

The mean and variance of the triangular distribution are:

$$\text{mean} = \frac{(a+m+b)}{3} \quad [3.7]$$

$$\text{variance} = \frac{(a^2+m^2+b^2-ma-ab-mb)}{18} \quad [3.8]$$

3.3.2.1.4. Exponential distribution

The exponential distribution is commonly used in conjunction with inter-arrival processes in simulation models because the arrival of entities in many systems has been empirically or theoretically shown to follow a random or Poisson process. This means that a random number of entities will arrive within a specific unit of time. Despite the randomness of the process, there will always be an average number of arrivals within the unit of time. The number of arrivals expected during the unit of time is randomly distributed around the mean value. In case a system has a random or Poisson distribution for the number of arrivals during a period, the time between arrivals follows an exponential distribution (Figure 3.5). This time between arrivals, known as the inter-arrival time, is characterized by the exponential distribution's single parameter. This value is the average of all the data. If the inter-arrival time conforms to an exponential distribution, there will be a greater number of observations with arrival times lower than the mean value and fewer observations with arrival times above the average value. Consequently, the number of observations continuously decreases as the arrival time increases.

For E(λ), λ > 0, the density function is as follows:

$$f(x) = \begin{cases} \lambda e^{-\lambda x} & \text{if } x > \lambda \\ 0 & \text{else} \end{cases} \quad [3.9]$$

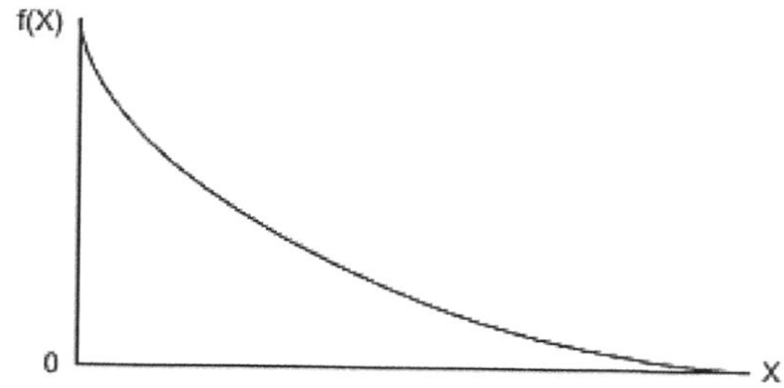

Figure 3.5. *Density function of the exponential distribution*

The statistical equations for the mean and variance of the exponential distribution are:

$$\text{mean} = \frac{1}{\lambda} \quad [3.10]$$

$$\text{variance} = \frac{1}{\lambda^2} \quad [3.11]$$

3.3.2.1.5. Poisson's distribution

The Poisson distribution is used to represent a random number of events that occur in a time interval (Figure 3.6). This distribution is characterized by a single parameter λ. Furthermore, this distribution is distinctive in the sense that both the mean and the variance are equal to λ. The probability of observing a particular value is determined by the formula:

$$p(x) = \frac{e^{-\lambda}\lambda^x}{x!} \quad [3.12]$$

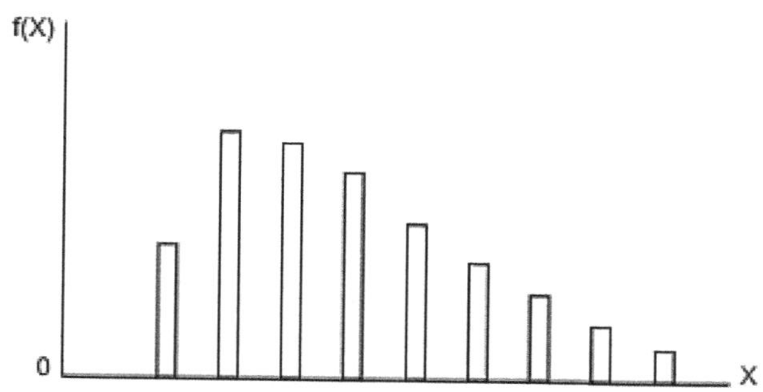

Figure 3.6. *Poisson's distribution density function*

Applications of the Poisson distribution in simulation encompass scenarios such as determining the number of arrivals within a specified time interval or calculating the count of entities in a batch.

3.3.2.1.6. Bernoulli's distribution

Bernoulli's distribution is used to model a random event with two possible outcomes: success or failure (Figure 3.7).

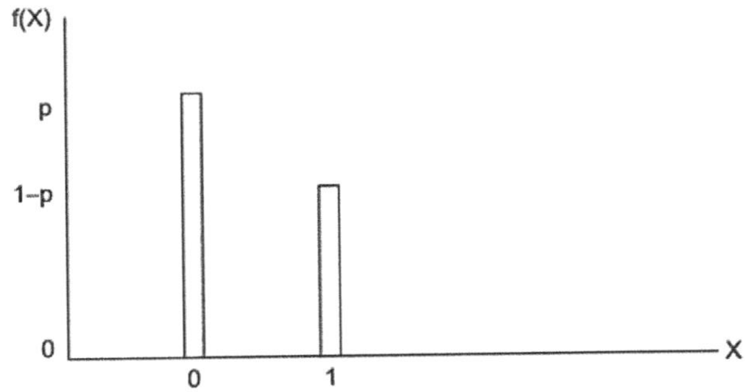

Figure 3.7. *Bernoulli's distribution density function*

The mean and variance of the Bernoulli distribution are as follows:

$$\text{mean} = p \quad [3.13]$$

$$\text{variance} = p(1-p) \quad [3.14]$$

where p = the fraction of successes and (1 – p) = the fraction of failures.

As examples, this Bernoulli's distribution finds use in simulation in:

– the pass or fail inspection processes;

– the classification of passengers in first class or economy class;

– the order of priority in an urgent or regular hospital.

In the case of the pass or fail inspection process, any part can only be in one of these two states. If 95% of parts pass inspection, the p-value is 0.95 and the 5% failure rate would be represented by 0.05. The process of screening passengers in first class or economy class does not really have a categorization into success or failure. In this case, it is simply the percentage of first class passengers and economy class passengers that must reach 1.0. A similar situation exists with urgent and regular priority orders. There is no success or failure associated with urgent or regular orders. They are simply two different order types with different probability percentages.

3.3.2.1.7. Weibull distribution

The Weibull distribution is often used to represent distributions that cannot have values less than zero (Figure 3.8). This situation frequently arises with symmetric distributions, such as the normal distribution, which represent service or process times. If the mean is small and the standard deviation is large enough, many observations accumulate on the left side of the distribution near 0. This leads to a non-symmetric distribution. The Weibull distribution is characterized by two parameters: the shape parameter (α) and the scale parameter (β). Depending on the values of these two parameters, the Weibull distribution can take forms ranging from that of the exponential distribution to that of the normal distribution. The probability function of the Weibull distribution is as follows:

$$f(x) = \begin{cases} \alpha\beta^{-\alpha}x^{\alpha-1}e^{-(x/\beta)^{\alpha}} & \text{if } x > 0 \\ 0 & \text{else} \end{cases} \quad [3.15]$$

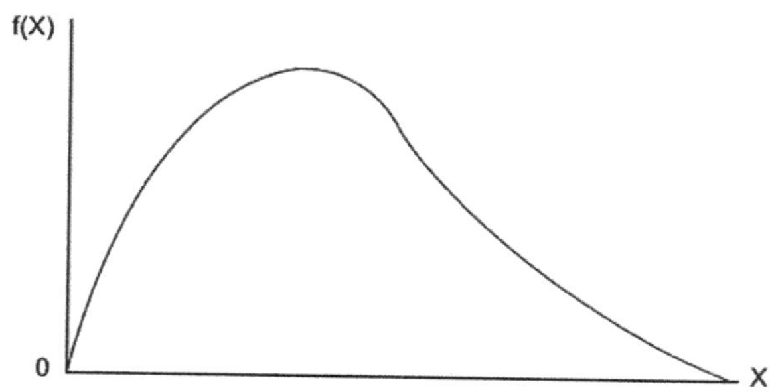

Figure 3.8. *Weibull density function*

The mean and variance of the Weibull distribution are represented mathematically as follows:

$$\text{mean} = \frac{\beta^2}{\alpha}\Gamma\left(\frac{1}{\alpha}\right) \quad [3.16]$$

$$\text{variance} = \frac{\beta^2}{\alpha}\left\{2\Gamma\left(\frac{2}{\alpha}\right) - \frac{1}{\alpha}\left[\Gamma\left(\frac{1}{\alpha}\right)\right]^2\right\} \quad [3.17]$$

where α and t β are the shape and scale parameters, respectively, and Γis given by:

$$\Gamma = \int_0^{\infty} x^{\alpha-1}e^{x}dx \quad [3.18]$$

3.3.2.2. *Non-existent data with partial knowledge*

This situation applies to systems which do not yet exist or for which it is impossible to have the desired data. In such cases, we must depend on estimates provided by operators, designers or even equipment suppliers.

3.3.2.2.1. Data available: the average M

In such case, we can use:

– M directly as a constant value of the variable if the dispersion, the standard deviation, is small;

– an exponential distribution, if the dispersion is large. This distribution will then have parameter $\lambda = M$.

3.3.2.2.2. Data available: Min and Max values

In the case where only the minimum and maximum values are available, we can use:

– a uniform distribution of Min and Max parameters;

– a normal distribution centered around M if the data are centered around the mean M = (Min + Max)/2. From the range of the data, which is in fact the difference between these two values, we can calculate the standard deviation. Indeed, if the data are numerous, then the standard deviation σ= Range/6.

3.3.2.2.3. Data available: Min and Max values and the most probable value m

In this case, we can use a triangular distribution with parameters Min, m and Max.

3.3.2.3. *Data accessible for measurement*

Depending on the measured data, mathematical analysis methods can be applied, which are as follows:

– If empirical data are utilized directly, they are inputted in the form of cumulative empirical distributions often represented as frequency histogram with observations grouped into classes with a number of classes = $\sqrt{\text{number of observations}}$).

– If we want to make adjustments of the data to theoretical distributions, we must:

- select a distribution based on its shape and that of the data histogram;
- estimate its parameters;

- perform a hypothesis test to investigate the fit of these data to the probability distribution.

3.3.2.3.1. Hypothesis statements

The null hypothesis is typically a statement that the input data could come from a specific theoretical distribution. Conversely, the alternative hypothesis is a statement that the input data could not have come from this specific theoretical distribution. In practice, the null and alternative hypotheses are condensed into the following format:

– H_0: distribution (parameter 1, parameter 2, etc.);

– H_1: no distribution (parameter 1, parameter 2, etc.).

For example, if service data are tested against a normal distribution, the mean and standard deviation must be available. For instance, if we hypothesize that the theoretical distribution follows a normal distribution with a mean of 5 and a standard deviation of 2, the assumption statement would be:

– H_0: normal (5, 2);

– H_1: not normal (5, 2).

3.3.2.3.2. Determining the significance level of the test

Establishing a confidence level for the test is crucial. For instance, if we aim to be 95% sure of the test results, the significance level will be 0.05. This significance level is commonly referred to as the "α" level. Other frequently used α levels are 0.01 and 0.10.

3.3.3. *Test of adequacy of the chosen distribution to the data*

Two tests can be used to examine the adequacy of the statistical distribution in relation to the system data: the chi-square test and the Kolmogorov–Smirnov (KS) test (Foucart 2000; Ycart 2002). The chi-square test assesses the goodness of fit by comparing actual and expected numbers of observations. Adequate data volume is essential for the application of the chi-square test. Typically, a minimum of 20 data points is required for the test. A more realistic minimum would be 30 data points. Below 30 data points, confidence in the test results diminishes, prompting consideration of alternative methods like the KS test.

3.3.3.1. *Chi-square test*

The procedure for conducting this test involves the following steps (Foucart 2000; Ycart 2002):

– Calculating the critical value of the chi-square distribution: the critical value is the limit between the non-significant and significant halves of the chi-square distribution. In other words, with a significance level of 0.05, the critical value represents the point where 95% of the distribution lies to the left of the critical value, while 5% of the distribution lies to the right of the critical value.

– Calculating the chi-square test statistic from the data: the test statistic is calculated by adding the squared differences between the observed number of data points and the expected number of data points divided by the expected number of points of data for each individual data cell. This process is more easily explained with the following formula:

$$\chi^2 = \sum_{i=1}^{n} \frac{(O_i - E_i)^2}{E_i} \qquad [3.19]$$

with:

- χ^2 = test statistic to calculate and compare to the critical value;
- O_i = number of observations of data points in the ith data cell;
- E_i = number of data points expected in the ith data cell;
- n = number of data cells.

– Comparison of the test statistic with the critical value in order to accept or reject the null hypotheses: in this step, the comparison involves comparing the test statistic, just calculated, with the previously determined critical value. If the test statistic turns out to be less than the critical value, the null hypothesis that the data could have come from this theoretical distribution cannot be rejected at the previously determined level of significance. Conversely, if the value of the test statistic is greater than the critical value, then the null hypothesis is rejected. This would mean that there is evidence that the data do not come from this theoretical distribution.

3.3.3.2. *Kolmogorov–Smirnov (KS) test*

This test is based on the following process (Foucart 2000; Ycart 2002):

– order the n observed values ($x_1,\ldots,x_n$) of a random variable X such that $x_1 \leq x_2 \leq \ldots \leq x_n$;

– calculate $F(x_1) = 1/n$, $F(x_2) = 2/n,\ldots$, $F(x_n) = 1$, which defines the staircase distribution function of F;

– calculate for each value of x_i the difference $|F(x_i) - F_0(x_i)|$, F_0 being the distribution function of the law whose adequacy to the data is tested;

– find the max value of this difference $D = \sup |F(x) - F_0(x)|$;

– find the critical value $d_{n,\alpha}$ in the KS law table;

– if $D < d_{n,\alpha}$, the adequacy hypothesis is accepted; otherwise, it is rejected.

3.3.3.3. *P-value*

The results of the chi-square and KS tests can be presented as “p”-values. The p-value is the largest value of error probability that allows the distribution to fit the data. In general, a higher p-value indicates a better fit. For example, if the p-value is greater than 0.10, we will not reject the null hypothesis of good fit at level = 0.10.

3.3.3.4. *Example*

A company counts the number of breakdowns that occur on machines over a period of 100 days and classifies them by workforce (Table 3.1).

Number of daily breakdowns	0	1	2	3	≥4
Observations	11	27	29	20	13

Table 3.1. *Number of daily breakdowns*

The following question arises: can we adjust these values to a Poisson distribution?

For a variable X following a Poisson distribution P(λ), it is known that E(X) = λ. Hence, λ is estimated using the empirical mean derived from the sample. This estimate is obtained by calculating the empirical mean

$$\overline{X} = \frac{1}{100}(11 \times 0 + 27 \times 1 + 29 \times 2 + 20 \times 3 + 13 \times 4) = 1.97 \qquad [3.20]$$

Let us round λ to 2 for simplification purposes. Now, we seek the theoretical numbers derived from a Poisson distribution with a parameter of 2. The probability for this distribution is defined by:

$$P(X = k) = e^{-\lambda}\frac{\lambda^k}{k!} \qquad [3.21]$$

We therefore find:

$P(X=0) = 0.135 \Rightarrow E_0 = 13.5$

$P(X=1) = 0.271 \Rightarrow E_1 = 27.1$

$P(X=2) = 0.271 \Rightarrow E_2 = 27.1$

$P(X=3) = 0.180 \Rightarrow E_3 = 18.0$

Finally, we deduce, by complement to 100, that $E_4=14.3$.

We now calculate the χ^2 statistic:

$$\chi^2=\sum_{i=1}^{n}\frac{(O_i-E_i)^2}{E_i}=\frac{(11-13.5)^2}{13.5}+\frac{(27-27.1)^2}{27.1}+\frac{(29-27.1)^2}{27.1}+\frac{(20-18)^2}{18}+\frac{(13-14.3)^2}{14.3}=0.937 \quad [3.22]$$

Given that we have five classes, the number of degrees of freedom is 4. The chi-square table gives a critical value equal to 9.488 at the threshold $\alpha = 5\%$ (Appendix 1). As the calculated statistic χ^2 of 0.937 is significantly below this critical value of 9.488, we cannot reject the hypothesis H_0.

4

Results Analysis

4.1. Introduction

Each simulation project must be crowned by the phase of collecting and analyzing the results. This phase provides the main value added to the simulation user as it endeavors to understand the behavior of the system and generate predictive models. The main steps making up this phase are as follows:

– Step 1: choice of simulation length and design of replications if necessary. A good choice of the simulation length as well as a good design for the simulation replications allow the analyst to obtain the maximum statistical information.

– Step 2: estimation of performance measures. Statistics provide the data to calculate point estimates and/or confidence intervals for the system parameters of interest. Crucial issues in estimation are the size of the sample to be collected and the independence of the observations used to calculate the statistics.

– Step 3: analysis of the results. The statistical estimates are used in turn to understand the behavior of the system and generate performance predictions under various scenarios of different input parameters or operating scenarios. Experimenting with other system designs helps elucidate their relative merits and highlight their design compromises.

If the last two steps are common to any type of simulation model, the first step depends on the nature of this model. Simulation models can be classified into two broad categories based on their time horizon: terminating models and non-terminating models.

4.2. Determining the number of replications for a terminating system

The distributions of inputs to simulation models are generally probabilistic in nature. Input variability naturally results in some variation in output performance measures. It is then inappropriate to recommend any action based on the results of a single simulation run or replication. To reduce the risk of making an incorrect recommendation, it is necessary to run a number of replications of the simulation and then make recommendations based on all the results obtained. The natural question that then arises is "How many replications should be performed?".

The replication analysis process begins by selecting an initial number of replications. The statistics for this initial set of replications are below used to check whether or not additional replications are needed at a particular confidence level "α". If additional replications are necessary, they should be run and the summary statistics and replication formulas for the process should be recalculated and their results rechecked.

A smaller number of initial replications will increase the likelihood that this initial number will be insufficient and additional replications will be required. A greater number of initial replications will decrease the likelihood that this initial number of replications will be insufficient. However, more unnecessary initial replications will result in the increased time spent. As a reasonable compromise to this situation, it is advisable to initially run 10 replications. Once the results of these 10 replications have been collected, we must calculate the final number of replications to be performed. This calculation is based on the mean and standard deviation of the results of these 10 replications already carried out. The general mathematical formula for estimating the standard deviation "s" of "n" replications is as follows:

$$s = \sqrt{\frac{\sum_{i=1}^{n} \bar{x}_i - \bar{x}}{n-1}} \qquad [4.1]$$

with:

– $\bar{x}_i$ the average of the results of a replication;

– $\bar{x}$ the average of the n replications.

Based on the obtained value of "s", the confidence interval (CI) relating to these results can be calculated through the following formula:

$$CI = t_{1-\alpha/2,n-1}\, s/\sqrt{n} \qquad [4.2]$$

where "t" is Student's law for the degrees of freedom "1 – α /2" and "n–1", and "n" is the initial number of replications previously chosen.

The CI can be used to calculate the optimal number of repetitions required. Achieving this involves selecting a precision level that aligns to the study's objectives. Two general approaches exist for objectively determining the level of precision. The first method is based on an absolute comparison of the CI at a particular tolerance level. The second method is based on a relative value of the CI compared to the sample mean.

4.2.1. *Absolute precision*

In the absolute precision approach, it is imperative to determine a tolerable level of precision. This precision value should be in the same unit as the sample data. Adjusting the CI equation to achieve the desired level of precision is essential in this regard

$$\text{Desired absolute precision} = \text{CI} = t_{1-\alpha/2,n-1}\, s/\sqrt{n} \qquad [4.3]$$

The number of necessary replications will therefore be equal to:

$$n = (t_{1-\alpha/2,n-1}\, s/\text{Desired absolute precision})^2 \qquad [4.4]$$

4.2.2. *Relative precision*

The alternative approach to absolute precision is that of relative precision. For a robust statistical analysis, the CI should be relatively low compared to the sample mean. The commonly used value for the desired relative precision is 0.10. This means that the goal is for the CI to be 10% of the sample mean for the data:

$$\text{Desired relative precision} = \frac{t_{1-\alpha/2,n-1} s/\sqrt{n}}{\bar{x}} \qquad [4.5]$$

The number of necessary replications will therefore be equal to:

$$n = (t_{1-\alpha/2,n-1}\, s/\bar{x}.\, \text{Desired relative precision})^2 \qquad [4.6]$$

4.2.3. *Example*

Consider the result of the 10 simulation replications of a security control system (Table 4.1).

Replication	Average system time
1	10.80
2	11.96
3	10.47
4	10.70
5	10.80
6	11.35
7	11.04
8	10.68
9	11.15
10	11.00

Table 4.1. *Result of 10 simulation replications for a security control system*

The general mean of these 10 values is 11.00, and the sample standard deviation is 0.42. Calculating confidence interval yields CI = 0.30, knowing that $t_{1-0.05/2,10-1} = 2.26$. For an absolute precision of 0.20, the number of replications needed is n = 23.04. We then round the number of replications at n = 24. The results of these replications are summarized in Table 4.2.

Replication number	Average system time	Replication number	Average system time
1	10.80	13	10.94
2	11.96	14	10.88
3	10.47	15	10.78
4	10.70	16	11.04
5	10.80	17	10.68
6	11.35	18	10.99
7	11.04	19	10.99
8	10.68	20	10.57
9	11.15	21	10.92
10	11.00	22	11.04
11	11.02	23	10.91
12	10.66	24	10.90

Table 4.2. *Results of 24 simulation replications for a security control system*

The next step in the process is to verify that these 24 repetitions are indeed sufficient for the designated level of absolute precision. If the new dataset meets the absolute accuracy level, no further replications should be simulated. On the other hand, if the new dataset exceeds the absolute precision level, a new number of replications to run must be recalculated.

The new values of the general mean and standard deviation are 10.93 and 0.29, respectively. The value of the recalculated CI is 0.12. This value is less than the chosen precision level of 0.20. The number of 24 replications is therefore sufficient.

4.3. Determining the number of replications for a non-terminating system

The process of choosing the simulation length for non-terminating systems differs significantly from that for terminating systems. Non-terminating systems start only once and then continue for an indefinite period. The system may never cease operation or intermittently close and reopen in the same state. In all cases, it is considered to operate for an indefinite period.

The system undergoes a transient initial state before stabilizing and reaching a constant state known as the steady state. Even if the initial transient state may not hold a significant statistical importance, it still needs modeling. We face the decision between two different approaches:

1) start with the empty system;

2) start with the loaded system.

Starting with an empty system may take a long period of time before the system passes the initial transient state. The challenge lies in determining how to identify the duration corresponding to this initial transient state.

Alternatively, we can start the simulation with entities already present within the system. At the simulation start, the model is loaded, necessitating observation of the system to ensure that the model begins with the same number of entities and resource states.

4.3.1. *Determination of steady state*

Analysis of results from non-terminating systems should exclusively be performed only with data from steady-state models. Therefore, before conducting any analysis, it is imperative to eliminate the initial transient state, if present. As the simulation proceeds, the initial transient state can be identified by a series of observations displaying continuously increasing performance values in output measurements.

Various statistical methods exist for determining the transient period, but the most practical approach involves plotting the performance measure of interest throughout the simulation to identify when the steady state is reached. Additionally, several alternative methods can be applied for determining the duration of the transitional period.

4.3.1.1. *Graphical approach*

In the graphical approach, the goal is to visually identify the point at which the slope of the initial transient state approaches 0. At that point, the output performance measure has reached its steady state. Clearly, this qualitative approach is highly subjective and susceptible to influence by individual interpretation (Figure 4.1).

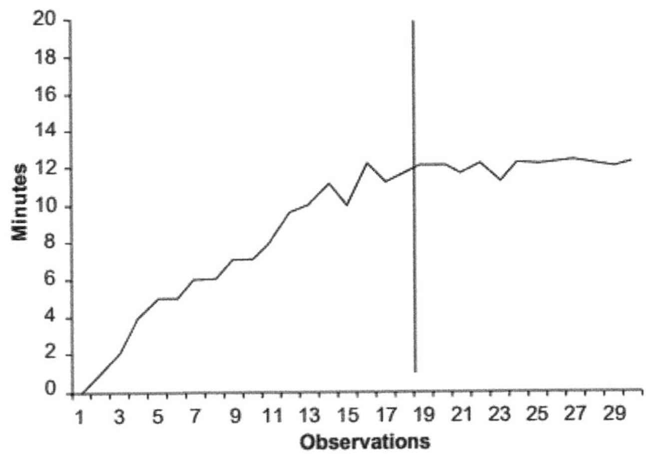

Figure 4.1. *Visual determination of the transitional period*

4.3.1.2. *The Welch method*

This is the simplest method of all the techniques used in determining the length of the transitional period. It is a graphical method that requires running several replications of the simulation. The steps in this method are as follows:

1) execute "n" replications of the simulation, each with a duration of "m", i.e. X_{ij} is the ith observation of the jth replication. Thus, i varies from 1 to m and j from 1 to n;

2) calculate all the averages $\overline{X}_i$ such that:

$$\overline{X}_i = \sum_{j=1}^{n} \frac{x_{ij}}{n} \text{ for } i = 1, 2, \ldots, m \quad [4.7]$$

3) calculate the motion average $\overline{X}_i(w)$ in order to eliminate high frequency oscillations in $\overline{X}_1, \overline{X}_2, \ldots, \overline{X}_m$. "w" is a positive integer, called the window, and its selection is at the discretion of the user. However, it must be less than or equal to m/4:

$$\overline{X}_i(w) = \frac{\sum_{s=-w}^{w} \overline{X}_{i+s}}{2w+1} \text{ for } i = w + 1, w + 2, \ldots, n - w \quad [4.8]$$

$$\overline{X}_i(w) = \frac{\sum_{s=-(i-1)}^{i-1} \overline{X}_{i+s}}{2i-1} \text{ for } i = 1, 2, 3, \ldots, w \quad [4.9]$$

4) draw the graph of $\overline{X}_i(w)$ for $i = 1, 2, 3, \ldots, n - w$ and find the value of i beyond which it $\overline{X}_i(w)$ begins to converge.

4.3.1.3. *The rule of crossing means*

If the number of crossing means exceeds a pre-specified value, the steady state can be considered to have been reached. The length of the transitional period can be estimated using the following steps:

1) generate the simulation results $x_1, x_2, \ldots, x_m$;

2) calculate w_j such that:

$$w_j = \begin{cases} 1, \text{if } \bar{x}_j > \bar{x}_m \text{ and } \bar{x}_{j+1} < \bar{x}_m \text{ or } \bar{x}_j < \bar{x}_m \text{ and } \bar{x}_{j+1} > \bar{x}_m \\ 0 \qquad\qquad \text{else} \end{cases} \quad [4.10]$$

where $j = 1, 2, \ldots, m - 1$ and $\bar{x}_m = \frac{1}{m}\sum_{j=1}^{m} x_j$;

3) calculate the numbers of crossings of the means $\Omega_1, \Omega_2, \ldots, \Omega_m$ until Ω_l is equal to the value prespecified by the user. The recommended value is 3. The numbers of crossings of the means are given by:

$$\Omega_m = \sum_{j=1}^{m-1} w_j \quad [4.11]$$

4.3.2. *Long simulation*

Long simulations are about finding a compromise between doing one simulation over a long period and doing several short-term simulations.

The long-duration simulation method includes only a single, very long simulation which is suitably divided, after elimination of the transient period, into "n" periods, each of which is treated as an independent simulation (Figure 4.2). Care must be taken in selecting the size "m"; longer periods can generate independent observations that can be considered independent simulations.

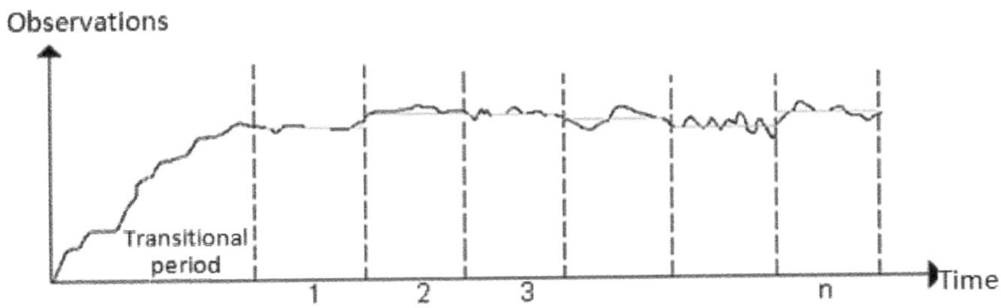

Figure 4.2. *Period average method*

4.4. Statistical analysis of results

4.4.1. *Simple comparison between two simulation models*

For simple comparisons of two models, either a hypothesis test or a confidence interval can be used. Hypothesis testing involves deciding whether to accept or reject the null hypothesis typically assuming that there is no difference between the two models. Conversely, the alternative hypothesis posits that there is a difference between the two models.

The confidence interval approach is a modification of the corresponding hypothesis test, establishing the range where the difference between the two models should normally be observed. Typically, a 95% confidence interval is employed. When the models are statistically similar, the difference between the mean replication values is expected to be zero. Thus, if the confidence interval includes 0, it statistically confirms that the two models are identical. On the other hand, if 0 falls outside the confidence interval, the two models are considered statistically different. There are two basic types of confidence interval approach to determining the difference between simulation models.

4.4.1.1. *Welch confidence interval approach*

The Welch confidence interval approach adopts a robust philosophy with respect to assumptions about data characteristics. With this approach, there is no need to be concerned about whether the two datasets possess similar variances.

After calculating the summary statistics of the mean and standard deviation for each dataset, the degrees of freedom estimator "dof" must be calculated using the following formula:

$$\text{dof} = \frac{\left[s_1^2/n_1+s_2^2/n_2\right]^2}{\left[s_1^2/n_1\right]^2/(n_1-1)+\left[s_2^2/n_2\right]^2/(n_2-1)} \quad [4.12]$$

with:

– dof: degree of freedom;

– s_1^2: the variance of the first simulation model;

– s_2^2: the variance of the second simulation model;

– n_1: the size of the observations of the first model;

– n_2: the size of the observations of the second model.

The number of degrees of freedom calculated in this way will probably not be an integer. In such cases, it will need to be rounded down.

The Welch confidence interval can now be calculated with the following formula:

$$\bar{x}_1 - \bar{x}_2 \pm t_{ddl,1-\alpha/2}\sqrt{\frac{s_1^2}{n_1}+\frac{s_2^2}{n_2}} \quad [4.13]$$

with:

– $\bar{x}_1$: the average of the responses of the first model;

– $\bar{x}_2$: the average of the responses of the second model;

– $t_{ddl,1-\alpha/2}$: the value obtained from the Student table.

The translation of the confidence interval approach, as discussed previously, only involves the relationship between the value 0 and the confidence interval. If the confidence interval covers the value 0, then there is no significant difference between the two simulation model alternatives. Conversely, if the confidence interval does not cover 0, then there is a statistically significant difference between the two simulation models.

Example: The following data represent overhead times in seconds for two different operating policies for a security control system airport. Solution A consists of two ticket checkers, two X-ray machines and two metal detectors. Alternative B consists of three ticket inspectors, two X-ray machines and a metal detector. For 10 observations of measurements for each alternative, the means and standard deviations of alternatives A and B are 39.58 seconds and 1.53 seconds and 45.60 seconds and 2.38 seconds, respectively.

To calculate the Welch confidence interval at a significance level α of 0.05, we start by estimating the degree of freedom:

$$\text{dof} = \frac{\left[1.53^2/10+2.38^2/10\right]^2}{[1.53^2/10]^2/(10-1)+[2.38^2/10]^2/(10-1)} = 15.34 \qquad [4.14]$$

Since we want to have as wide a confidence interval as possible, we round the dof to 15. This results in a t-value of 2.131. The actual calculation of the Welch confidence interval is:

$$39.58 - 45.60 \pm 2.131\sqrt{\frac{1.53^2}{10}+\frac{2.38^2}{10}}$$

$$= -6.02 \pm 1.92 \text{ or even } [-7.93; -4.11] \qquad [4.15]$$

The confidence interval does not include the value of 0 with a significance α level of 0.05. Therefore, it can be concluded that the two alternatives are statistically different, because solution A has a lower overhead of 39.58 seconds compared to the overhead of alternative B with an overhead of 45.60 seconds. Under normal circumstances, alternative A is the most recommended.

4.4.1.2. *Paired* t-*test confidence interval approach*

The paired *t*-test is used when the two models exhibit some sort of natural matching (Foucart 2000; Ycart 2002). In this approach, we first calculate a new variable based on the difference between the means of the simulation replications of the two alternatives:

$$\overline{X}_{1i} - \overline{X}_{2i} = Z_i \qquad [4.16]$$

with:

– $\overline{X}_{1i}$ being the average of the ith replication of the first alternative;

– $\overline{X}_{2i}$ being the average of the ith replication of the second alternative;

– Z_i being the difference in means for the ith replication.

With 10 replications, we obtain 10 Z values for each pair of replication means. Then, the mean and the standard deviation of the new variable Z must be calculated. Finally, the confidence interval must be calculated based on the following formula:

$$\bar{Z} \pm t_{\alpha/2,n-1}\frac{s}{\sqrt{n}} \qquad [4.17]$$

where:

– $\bar{Z}$ is the average of the Z_i values;

– $t_{\alpha/2,n-1}$ is the value obtained from the Student table for $\alpha/2$ and n–1 degrees of freedom;

– s is the standard deviation of the Z_i values;

– n is the number of pairs of differences of replication means.

For example, if we had obtained the data in the example from two alternatives of the same model, the calculations would be made at a significance level α of 0.05 (Table 4.3).

Replication	Alternative A	Alternative B	Difference Z
1	38.88	42.44	–3.56
2	43.06	44.53	–1.47
3	37.69	46.55	–8.86
4	38.52	48.46	–9.94
5	38.88	48.92	–10.04
6	40.86	48.46	–7.60
7	39.74	45.36	–5.62
8	38.45	44.53	–6.08
9	40.14	42.84	–2.70
10	39.60	43.92	–4.32

Table 4.3. *Result of the difference of replications*

The Z mean is –6.02 and its standard deviation is 3.04. The t value for 0.05/2 with nine degrees of freedom is 2.262. The CI of the Z is then:

$$-6.02 \pm 2.262\frac{3.04}{\sqrt{10}} = -6.02 \pm 2.29 \text{ or even } [-8.31; -3.73] \qquad [4.18]$$

Since the CI does not include the value of 0, there is statistical evidence to support that the two alternatives are statistically different at a significance level α = 0.05. Given that the average of alternative A is 39.58 seconds and that of alternative B is 45.60 seconds, and given our aim to minimize the studied performance, the preferable choice would typically be alternative A.

4.4.2. Comparison of three or more simulation models

Three or more comparisons involve a two-step procedure. In the first step, analysis of variance (ANOVA) is used to determine if there is a statistically significant difference between one and more of the means of the different models. If none of the means are statistically different, the analysis can be stopped. On the other hand, if one or more means are statistically different, it is crucial to identify which means are involved. Unfortunately, ANOVA alone does not provide this information. In order to determine which means are statistically different from others, the multiple ranges Duncan test can be used (Foucart 2000; Ycart 2002).

4.4.2.1. *Analysis of variance (ANOVA)*

ANOVA can be used to determine whether the average performance of one or more alternatives is statistically different from the others at a given α significance level (Foucart 2000; Ycart 2002). One or more of the averages might actually be statistically significantly better or worse than the rest of the alternatives. The analysis of variance is based on a ratio of the variance between different alternatives divided by the variance within the different alternatives. If the variance between the alternatives is large and the variance within the different alternatives is small, the ratio takes on a relatively high value. Conversely, if the variation between the alternatives is small and the variance within the different alternatives is large, the ratio is low. When the ratio is high, it is more likely that one or more of the alternatives will be statistically different from the others.

To begin the ANOVA process, each of the means of individual replications of the different alternatives must be provided. The average value of the replications for each alternative and the single general average of all the replications must also be provided. With these statistics, the following calculations are necessary:

– calculation of the sum of total squares;

– calculation of the sum of squares between groups;

– calculation of the sum of squares within groups;

– calculation of the mean of squares between groups;

– calculation of the mean of squares within groups;

– calculation of the F statistic;

– comparison of the F statistic to a critical F value.

4.4.2.2. *Calculating the sum of total squares*

The total sum of squares, SC_T, is the sum of the difference between the mean of each individual replication and the grand mean squared. This term is presented by the following formula:

$$SC_T = \sum_{i=1}^{k} \sum_{j=1}^{n} \left(x_{ij} - \bar{\bar{x}}\right)^2 \qquad [4.19]$$

where:

- SC_T is the total sum of squares;
- k is the number of different alternatives;
- n is the number of replications for each alternative;
- x_{ij} is the average of a single replication for an alternative;
- $\bar{\bar{x}}$ is the total average of all replications.

4.4.2.2.1. Calculation of the inter-group sum of squares

The inter-group sum of squares, SC_{inter}, is calculated by adding the difference between the means of the individual replications and the alternative mean squared. The SC_{inter} is calculated using the following formula:

$$SC_{inter} = \sum_{i=1}^{k} n(\bar{x}_i - \bar{\bar{x}})^2 \qquad [4.20]$$

where:

- SC_{inter} is the sum of squares between groups;
- k is the number of different alternatives;
- n is the number of replications for each alternative;
- $\bar{x}_i$ is the average of all replications for an alternative;
- $\bar{\bar{x}}$ is the total average of all replications.

4.4.2.2.2. Calculation of the intra-group sum of squares

The intra-group sum of squares, SC_{intra}, must be calculated. Each of these terms can be individually calculated by adding the squared differences of each individual replication mean and its alternative mean based on the following formula:

$$SC_T = SC_{inter} + SC_{intra} \qquad [4.21]$$

By rearranging this formula, the sum of squares can be obtained using what we already know and we can save a considerable amount of calculations:

$$SC_{intra} = SC_T - SC_{inter} \qquad [4.22]$$

4.4.2.2.3. Calculation of the inter-group mean of squares

The inter-group mean of squares, MC_{inter}, is calculated by dividing the sum of the inter-group squares by the number of degrees of freedom associated with the alternatives.

The number of degrees of freedom is the number of alternatives reduced by 1. The equation associated with this calculation is the following:

$$MC_{inter} = \frac{SC_{inter}}{k-1} \qquad [4.23]$$

where:

– MC_{inter} is the mean of the squares between groups;

– SC_{inter} is the sum of squares between groups;

– k is the number of alternatives.

4.4.2.2.4. Calculation of the intra-group mean of squares

The intra-group mean of squares, MC_{intra}, is calculated by dividing the inter-group sum of squares by the number of degrees of freedom associated with this sum of squares. The equation associated with this calculation is as follows:

$$MC_{intra} = \frac{SC_{intra}}{k(n-1)} \qquad [4.24]$$

where:

– MC_{intra} is the mean of the squares within groups;

– SC_{intra} is the sum of squares within groups;

– k is the number of alternatives;

– n is the number of replications for each alternative.

4.4.2.2.5. Calculation of the F statistic

The next step in the analysis of variance is to calculate the F statistic. This statistic is the ratio between the inter-group mean squares and the intra-group mean squares. This amounts to comparing the variance between alternatives to the variance within alternatives. The F statistic can be calculated based on the following equation:

$$F = \frac{MC_{inter}}{MC_{intra}} \quad [4.25]$$

where:

– F is the F statistic;

– MC_{inter} is the mean of the squares within groups;

– MC_{inter} is the mean of the squares between groups.

4.4.2.2.6. Compare the F statistic to the critical value

The final step is to compare the F statistic to the critical F value obtained using an F distribution table. The necessary parameters to determine the critical F value are the level α, the number of degrees of freedom in the numerator and the number of degrees of freedom in the denominator. The first number of degrees of freedom is the same number that was used to divide the $SC_{inter.}$ As for the second number of degrees of freedom, it is the same as that which was used to divide the $SC_{intra.}$ These degrees of freedom are as follows:

$$dof_{numerator} = k - 1 \quad [4.26]$$

$$dof_{numerator} = k(n - 1) \quad [4.27]$$

The calculated F-test statistic is finally compared to the critical F-value obtained from the F-distribution table. If the test statistic is greater than the critical value, then at least one of the alternatives is statistically different from the others. Conversely, if the test statistic is less than the critical value, then all alternatives are statistically similar.

Example: It is very unusual to perform the preceding ANOVA calculations by hand. Indeed, there are several software programs, such as Minitab, which easily carry out this calculation. Table 4.4 summarizes the result of an ANOVA analysis applied to the simulation results of the operating policies for the safety control system of an airport. A complete experimental design was applied to the Ticked, X-ray and Metal factors with their two levels each. This plan therefore includes eight experiments each replicated 67 times. Analysis of the results indicates that at one level of significance of 0.05, each of the main factors (Ticket, X-ray and Metal) is statistically significant (Table 4.4). This means that changing the level of any of these factors will have a statistically significant impact on the time it takes for an air passenger to go through screening. The interactions between the factors do not present significant effects (Appendix 2).

Source	dof	SC	MC	F	P
Ticket	1	906.48	906.48	4.83	0.028*
X-ray	1	92,370.60	92,370.60	492.19	0.000*
Metal	1	1,944.36	1,944.36	10.36	0.001*
Ticket x X-ray	1	29.52	29.52	0.16	0.692
Ticket x Metal	1	16.56	16.56	0.09	0.766
X-ray x Metal	1	662.76	662.76	3.53	0.061
Ticket x X-ray x Metal	1	15.84	15.84	0.08	0.772
Error	528	99,091.08	187.56		
Total	535	195,037.20			

* $p < 0.05$; therefore, significant effect.

Table 4.4. *ANOVA analysis result (example)*

4.4.2.3. *Multiple range Duncan test*

If the analysis of variance reveals statistically significant differences among the means, a subsequent multiple range Duncan test becomes imperative. This test discerns which means diverge from others with statistical significance at a chosen confidence level. Consequently, Duncan's test provides additional insights beyond ANOVA to make informed recommendations. Obviously, performing a Duncan test when the ANOVA results do not indicate significant statistical difference between the means would be futile.

The general concept of the Duncan test entails determining the least significant interval value for a given set of adjacent means. These means are defined as any series arranged in order. This series can consist of pairs, triplets or more consecutive averages. The core principal of the Duncan test is that any set of adjacent means with a maximum–minimum deviation exceeding the calculated critical value of the least significant deviation is statistically insignificant. In other words, if the difference between the maximum and minimum values of an adjacent set is less than the critical value, then all means of the adjacent set can be considered statistically similar.

Since adjacent sets can involve two, three, four or more sets of adjacent means, the Duncan test can involve quite a large number of calculations. Fortunately, if it has been proven that a large set of adjacent means is statistically similar, then all small sets of adjacent means within the large set will also be statistically similar. This means, for example, that if a set of four adjacent means is found to be statistically similar, all combinations of two or three sets of adjacent means are also statistically similar.

The Duncan multiple range test procedure includes the following steps:

1) sorting the replication means for each alternative in ascending order from left to right;

2) calculating the value of the least significant interval for all possible sets of adjacent means;

3) comparing each set of possible adjacent means with the value of the corresponding least significant interval in descending order with respect to the size of the set;

4) marking the non-significant interval.

4.4.2.3.1. Calculating the value of the least significant interval

To ascertain the least significant value of the interval for each set of adjacent means, a total of n−1 calculations are required (for p=2 to n), where "n" denotes the number of alternatives. The smallest set will consist of two means, while the largest set will match the number of alternatives. Additionally, it is crucial to designate a significance level, denoted as α, for calculating the least significant interval. Typically, for a confidence level of 95%, α is set to 0.05. The calculation process for determining the values of the least significant interval is a two-step process. In the first step, a value derived from the root mean square error of the replication means and the number of replications within a single alternative must be calculated based on the following formula:

$$s_{\bar{x}} = \sqrt{\frac{MC_e}{n}} \qquad [4.28]$$

where:

– MC_e is the mean square error of the means of the replications;

– n is the number of replications in a single alternative.

The value of the mean square error of the replication means can readily be obtained from the ANOVA statistical report. Particular care should be taken to ensure that this specific "n" represents the number of replications within a single alternative, not the number of alternatives or the total number of replications analyzed in the study. In the subsequent step of calculating the least significant interval, a different "n" will be used.

Once the Duncan standard deviation of the replication means has been calculated, it will be multiplied by a specific multiplier sourced from the multiple range Duncan test tables to derive the least significant range values. The formula to determine the value of the least significant interval is as follows:

$$R_p = s_{\bar{x}} r_p \qquad [4.29]$$

where:

– $s_{\bar{x}}$ is the Duncan standard deviation of the replication means;

– r_p is the Duncan multiplier of the replication means for a given significance level, set size and degrees of freedom;

– p is the size of the set of adjacent means.

The multi-range Duncan multiplier table requires input parameters, including the chosen significance level, the size of the adjacent set of means and the number of degrees of freedom. This number aligns with those used to calculate the mean square error of the alternative replication means, readily accessible in the ANOVA results (Appendix 3).

The above calculations will give a total of n–1 values for the least significant interval using p values from 2 to n. Thus, with four different alternatives, we obtain three values for the sizes of sets of adjacent means of 2, 3 and 4.

4.4.2.3.2. Comparison of adjacent means

The next step involves comparing the values of the least significant interval to each set of adjacent means. Optimal practice suggests commencing with the largest set of adjacent means and processing towards the smallest set of adjacent means. This approach significantly reduces the number of required comparisons. The first comparison will be on the set of adjacent means which corresponds to the number of alternatives. If the range or difference between the largest mean and the smallest mean falls below the least significant value of the range, neither mean is statistically different from the others. Importantly, the interval of the largest set of adjacent means must always be greater than the value of the corresponding least significant interval. This necessity arises because the analysis of variance indicates that at least one mean of all alternatives is different. Conversely, not all adjacent means can be statistically similar.

Hence, the interval of the largest set of adjacent means must exceed the value of the least significant interval. If the interval of the largest set of adjacent means is smaller than the value of the least significant interval, then some kind of error has occurred. This discrepancy may stem from either the result of the analysis of variance suggesting statistical similarity among all means, or an error calculation during the process.

The next step is to compare the subsequent largest set of adjacent means. This set will include a total of two different sets of adjacent averages. The first set spans from the smallest average to the next largest average, while the second comprises the smallest average to the largest average. This means that if we had a total of four alternatives, the first set would encompass the three alternatives 1, 2 and 3 and the second set would contain alternatives 2, 3 and 4.

If one of these sets has an interval less than the value of the least significant interval for that set size, it indicates no significant difference between the means of that set. Consequently, any smaller set of adjacent means is also statistically similar, warranting no further analysis. Conversely, if one of these sets has an interval that is larger than the value of the least significant interval for that set of size, it denotes a statistically significant difference between the means of this set.

The analysis process should continue with this set of averages. The averages must be divided into a set of adjacent averages of immediately smaller size. For instance, if the interval of the set of three adjacent means for alternatives 1, 2 and 3 exceeds the least significant interval value, attention shifts to adjacent means of size two.

The first set of two adjacent means would involve alternatives 1 and 2. Subsequently, the second set of two adjacent means would comprise alternatives 2 and 3. This arrangement is based on the understanding that among these three adjacent means, the different mean must come from either the set of alternatives 1 and 2, or of the set of alternatives 2 and 3. If it turns out that the interval for the first set with alternatives 1 and 2 exceeds the least significant interval, it indicates the existence of a statistically significant difference between alternatives 1 and 2.

Conversely, if this same set does not exceed the least significant interval, it suggests no statistically significant difference between alternatives 1 and 2. Similarly, the interval for all alternatives 2 and 3 may or may not exceed the value of the least significant interval.

Example: the airport security screening system example illustrates the process of performing the multiple range Duncan test. First, the average system times for each of the alternatives are listed in ascending order (Table 4.5).

Order	1	2	3	4	5	6	7	8
Alternative	D	A	B	C	H	F	E	G
Ticket	3	2	3	2	3	3	2	2
X-ray	2	2	2	2	1	1	1	1
Metal	2	2	1	1	2	1	2	1
Average	35.07	39.10	46.50	50.56	89.51	90.59	96.97	97.70

Table 4.5. *Sort order of results of airport checkpoint simulation alternatives*

The mean squared error can be taken from the ANOVA example with the airport security screening system and used to calculate the Duncan standard deviation of the replication means:

$$s_{\bar{x}} = \sqrt{\frac{187.56}{67}} = 1.67 \qquad [4.30]$$

The Duncan test multiplier table values for r between 2 and 8 at $\alpha = 0.05$ and 120 degrees of freedom are shown in Table 4.6.

P	2	3	4	5	6	7	8
R	2.80	2.95	3.04	3.12	3.17	3.22	3.25

Table 4.6. *The values of the "r" multipliers*

It should be noted that the number of degrees of freedom for error in the analysis of variance table was 528. The closest number available in most Duncan multiple range tables is 120 degrees of freedom.

The r values are then multiplied by the standard deviation of the replication means given in equation [4.30] to calculate the least significant interval values.

P	2	3	4	5	6	7	8
R	4.68	4.93	5.08	5.21	5.29	5.38	5.43

Table 4.7. *The new values of the "r" multipliers multiplied by the standard deviation*

For the set of eight adjacent means, the least significant value of the interval is 5.43. The interval between the largest and smallest averages is:

$$97.70 - 35.07 = 62.63$$

This value is much higher than the least significant interval value of 5.43 for eight adjacent means. This means that at least one of the averages is statistically different from the others.

The next step is to examine the sets of alternatives with combinations of seven adjacent averages. The first is the combination D to E and the second is the combination A to G. The value of the least significant interval for seven adjacent means is 5.38. The calculated intervals of the two sets are:

1) 96.97 – 35.07 = 61.90 for the combination going from D to E;

2) 97.70 – 39.10 = 58.60 for the combination from A to G.

These two values exceed 5.38, so there is at least one statistically different mean in these two sets.

Following the same procedure, we have a total of three sets with six adjacent averages. These are D to F, A to E and B to G. The least significant value of the interval is 5.29 for six adjacent means. The calculated intervals of these averages are:

1) 90.59 – 35.07 = 55.52 for the combination going from D to F;

2) 96.97 – 39.10 = 57.87 for the combination from A to E;

3) 97.70 – 46.50 = 51.20 for the combination going from B to G.

Again, all of these ranges exceed the least significant range value of 5.29. Likewise, this means that at least one of the six means in each of these sets is statistically different from the others.

We then examine the sets with five adjacent means. The value of the least significant interval is 5.21. The sets of five adjacent means are D to H, A to F, B to E and C to G:

1) 89.51 – 35.07 = 54.44 for the combination going from D to H;

2) 90.59 – 39.10 = 51.49 for the combination going from A to F;

3) 96.97 – 46.50 = 50.47 for the combination going from B to E;

4) 97.70 – 50.56 = 47.14 for the combination going from C to G.

Once again, none of the four sets of five adjacent means is less than the least significant value of the interval. This means that in each of these four sets, at least one of the means is statistically different from the others.

With sets of four adjacent means, we have a least significant interval value of 5.08. Our sets of adjacent means are D to C, A to H, B to F, C to E and H to G. The interval values for these sets are:

1) 50.56 – 35.07 = 15.49 for the combination going from D to C;

2) 89.51 – 39.10 = 50.41 for the combination going from A to H;

3) 90.59 – 46.50 = 44.09 for the combination going from B to F;

4) 96.97 – 50.56 = 46.41 for the combination going from C to E;

5) 97.70 – 89.51 = 8.19 for the combination going from H to G.

Again, all values in the interval exceed the least significant value in the interval of 5.08. Since these sets of adjacent means are not statistically similar, we must move to sets of three adjacent means.

With three adjacent averages, a total of six different sets can be created. The least significant value of the interval for three adjacent means is 4.93. The sets of three adjacent means are D to B, A to C, B to H, C to F, H to E and F to G. The values of the interval are:

1) 46.50 – 35.07 = 11.43 for the combination going from D to B;

2) 50.56 – 39.10 = 11.46 for the combination going from A to C;

3) 89.51 – 46.50 = 43.01 for the combination going from B to H;

4) 90.59 – 50.56 = 40.03 for the combination going from C to F;

5) 96.97 – 89.51 = 7.46 for the combination going from H to E;

6) 97.70 – 90.59 = 7.11 for the combination going from F to G.

Although a few of these ranges approach 4.93, all ranges still exceed the value of the least significant range. For at least one other level of adjacent means, we see that there is at least one mean which is statistically different from the others. We then move to the last level of adjacent averages.

The value of the least significant interval for two adjacent means is 4.68. As there must be at least two adjacent ones to perform the multiple range Duncan test, a certain type of result is expected with this iteration. The sets of adjacent means are D to A, A to B, B to C, C to H, H to F, F to E and E to G. These interval values are:

1) 39.10 – 35.07 = 4.03 for the combination going from D to A;

2) 46.50 – 39.10 = 7.40 for the combination going from A to B;

3) 50.56 – 46.50 = 4.06 for the combination going from B to C;

4) 89.51 – 50.56 = 38.95 for the combination going from C to H;

5) 90.59 – 89.51 = 1.08 for the combination going from H to F;

6) 96.97 – 90.59 = 6.38 for the combination going from F to E;

7) 97.70 – 96.979 = 0.73 for the combination going from E to G.

The adjacent sets of means for D to A, B to C, H to F and E to G have deviations less than the least significant deviation value of 4.68. This means that there is no statistical difference for each of these pairs of means. Conversely, the sets of adjacent means A to B, C to H and F to E have deviations greater than 4.68. These sets of means are statistically different. The following conclusions can be drawn:

– Alternatives D and A are identical.

– Alternatives B and C are identical.

– Alternatives H and F are identical.

– Alternatives E and G are identical.

The following additional conclusions can also be drawn based on pairs that are not statistically identical:

– Alternative D performs significantly better than all other alternatives except alternative A.

– Alternative A is significantly better than all other alternatives except alternative D.

– Alternative B is significantly better than all other alternatives except alternatives D, A and C.

– Alternative H is significantly better than alternatives E and G.

– Alternative F is significantly more efficient than alternatives E and G.

At first glance, it would appear that the system should be operated with alternative D. Although this alternative offers the highest level of performance, it might not be the best overall choice among the system configurations.

The conclusions of the multiple range Duncan test make more sense when considering the operational costs of the system. For instance, there is no difference in performance between alternative D with seven individuals and alternative A with six individuals. As there is no statistically significant difference between these two alternatives, the system could be operated more cheaply with only six individuals rather than seven. Alternatively, the results can inform the establishment of a particular staffing level. For example, if only five equally trained and qualified individuals were available to operate the security control system, alternatives C, F and E emerge as viable options. Given that solution C outperforms both F and E statistically, it stands out as the optimal choice of personnel allocation.

5

Presentation of Arena Software

5.1. Introduction

Arena is a flow simulation software based on the discrete event technique to simulate the dynamic behavior of systems in different fields. This software is based on the SIMAN language, which is a modeling language and simulation of discrete event systems. Moreover, Arena integrates many other modeling functions, such as graphic animation and statistical analysis of results and their optimization. Hence, it allows the user to carry out a complete simulation project through programming, verification, validation, interpretation of results, optimization, etc. In fact, Arena consists of three different elements. The main component of Arena is a graphics software package that uses the SIMAN simulation programming code. Arena's graphic base allows the user to quickly develop models and easily animate them. The second component of Arena is the input analyzer. This software package is used to adjust the input data to statistical laws. The final component of Arena is the output analyzer, which allows statistical analysis and comparison of simulation model output performance measures.

Arena is based on different individual structures called model blocks and simulation elements. These blocks control the flow of entities in the model while the simulation elements define the characteristics of the different components of the model.

The approach adopted in this chapter consists of the presentation of simulation blocks and elements. Using these structures helps the reader better understand how the system works.

5.2. Arena software user interface

Arena uses different operating system windows as an interface. This interface contains the normal menu bar and icon toolbars, a "project bar" on the left side of the screen, a "model window flowchart view" in the center of the screen and a "model window spreadsheet view" giving a view on the model data at the bottom of the screen. Figure 5.1 illustrates the interface of the Arena software (Arena 2018).

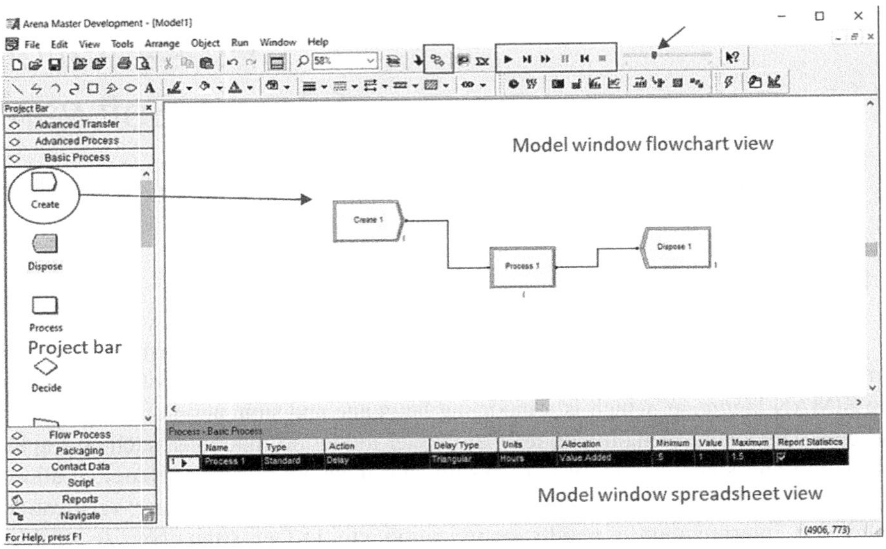

Figure 5.1. *Arena's simulation environment*

The menu bar contains the usual "File", "Edit", "View" and "Help" menus . Some of the most common functions are duplicated on the toolbar 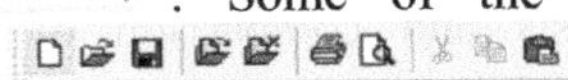. Besides these common functions, the toolbar also contains Arena-specific functions to connect simulation modules 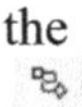, launch simulations 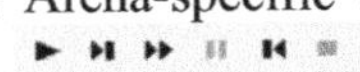, control simulation speed 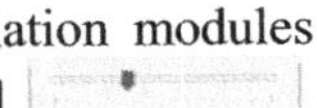 or even display variables and draw curves 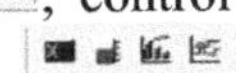.

The "Project Bar", located on the left of the screen, presents the different modules that can be used to build the simulation project. If a particular panel does not appear, it can be attached by right-clicking above the project toolbar.

The central model visualization screen, "flowchart view", serves to build the simulation model module by module. The model is built by dragging modules from the project toolbar, on the left of the screen, onto the model. The flow of the model is established by connecting the individual modules together through the connection icon .

The lower data screen, "spreadsheet view", is used to visualize and/or modify the parameters of the simulation modules. To increase the model visualization space, it will be useful to minimize the size of this data visualization screen.

Once the simulation model has been developed, it can be launched using the launch icon . It can also be temporarily stopped using the pause icon . This icon is to be used to completely stop the simulation and return the model to its initial state. The animation speed of the simulation model can be adjusted by moving the cursor of this icon to the right to increase it or towards the left to decrease it. These two icons can be used to animate the simulation model step by step.

5.3. The "basic process" panel

This section describes the flowchart and data modules that are most used to model systems (Arena 2018).

5.3.1. *Flowchart modules*

5.3.1.1. *"Create" module*

This module is considered as the starting point for entities in simulation models (Figure 5.2). The parameters of the "Create" module are:

– "Name": this is the name that the user must give to each module.

– "Entity Type": this is the type of the entity created by this module.

– "Time between arrivals": this is the inter-arrival time of the entities.

– "Type": entities can be created through planning or based on inter-arrival time. In the second case, this time can be constant or follow a statistical law.

– "Value": in the case where the inter-arrival time is constant, this value corresponds to the number of arrivals per unit of time. If this time follows a statistical law, the user must specify the characteristics of this law.

– "Units": this is the unit of time used.

– "Entities per Arrival": this is the number of arrivals per unit of time.

– "Max Arrivals": this is the maximum arrivals.

– "First Creation": this is the moment when arrivals begin.

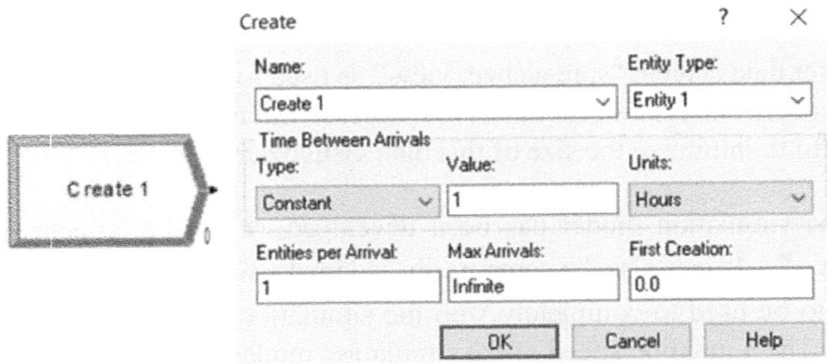

Figure 5.2. *"Create" module*

5.3.1.2. *"Dispose" module*

This module acts as an exit point for the entities in the simulation model (Figure 5.3). Once an entity has completed its journey through the system, there is no need to retain it further in the model. By removing the entity, program resources are freed for other purposes. Additionally, the "Dispose" module can record the statistics of the entity before it leaves the system. To do this, simply check the "Record Entity Statistics" option (Arena 2018).

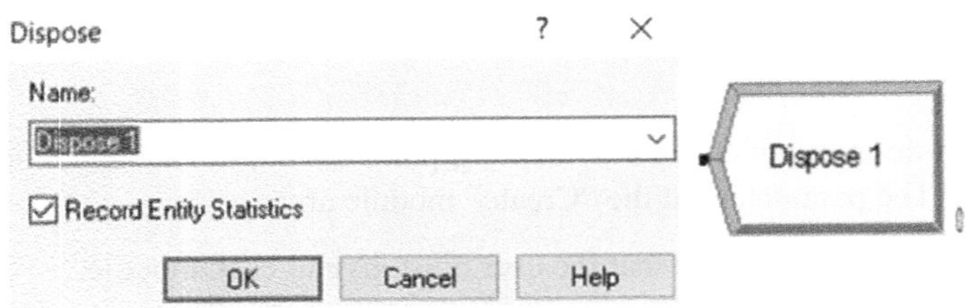

Figure 5.3. *"Dispose" module*

5.3.1.3. *"Process" module*

This module is designed as the main execution tool in a simulation model (Figure 5.4). The execution time allocated to the entity can be considered as value added, non-value added, transfer, wait or other. The associated cost will be added to the appropriate category (Arena 2018).

Different options for seizing and releasing resources associated with this module are available:

– "Delay" simply indicates that the entity will be delayed by this module without any resource constraints.

– "Seize Delay" indicates that one or more resources will be allocated in this module and a delay will occur, but the resource will be released later.

– "Seize Delay Release" indicates that one or more resources will be allocated, followed by a delay in the process, and then the allocated resource(s) will be released.

– "Delay Release" indicates that one or more resources have already been allocated and the entity will simply delay and release the specified resource(s).

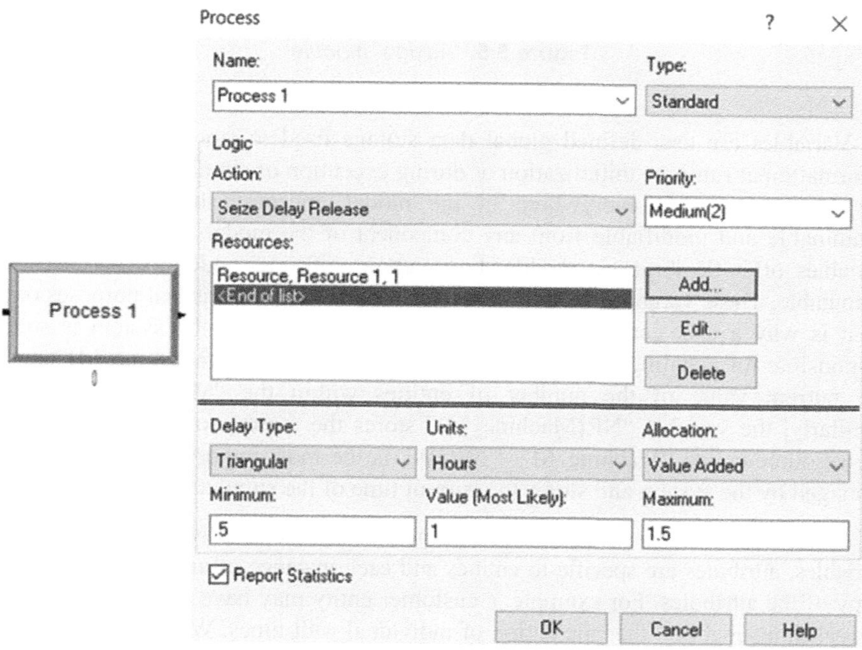

Figure 5.4. *"Process" module*

5.3.1.4. *"Assign" module*

This module can be used to assign particular values to global variables or entity attributes (Figure 5.5). Additionally, this module enables the assignment of a type or a different picture to the entity. Multiple assignments can be performed using a single "Assign" module (Arena 2018).

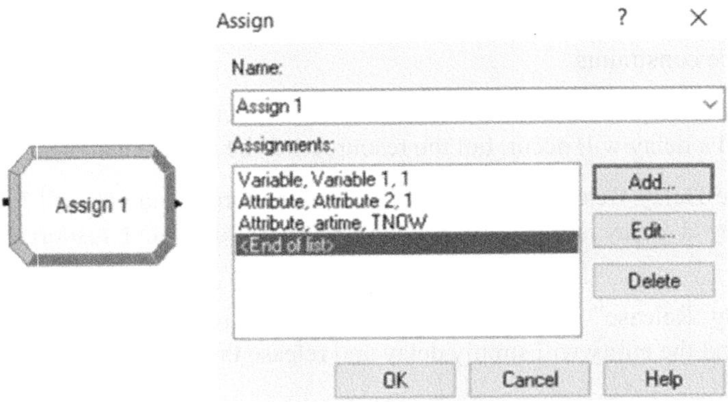

Figure 5.5. *"Assign" module*

Variables are user-defined global data storage used to store and modify state information at runtime, initialization or during execution of simulation. These global variables are visible everywhere in the model, meaning they are accessible, examinable and modifiable from any component of the model. While user-defined variables offer flexibility in modification, certain software-predefined variables are immutable. These varables are designated for observation or statistical purposes only. That is why a user cannot assign a new value to them and the system is solely responsible for apdating them. For example, the variable "NQ(Machine_M)" stores the current value of the number of entities within the "Machine_M" queue. Similarly, the variable "NR(Machine_M)" stores the number of occupied units of the resource called "Machine_M". "TNOW" is the most important variable that is managed by the system and stores the current time of the simulation.

Attributes are data storage objects associated by the user with entities. Unlike variables, attributes are specific to entities and each instance of an entity has its own copy of the attributes. For example, a customer entity may have an attribute such as arrival time, enabling the calculation of individual wait times. When arrivals consist of multiple customer types, the arrival type can also be stored as an attribute of each customer entity to allow separate statistics to be collected for each customer type.

5.3.1.5. *"Decide" module*

The "Decide" module allows decision-making within the simulation model (Arena 2018). It offers versatile options for making decisions based on one or more conditions depending on various factors such as entity type or probabilities, like 50% true and 50% false. Conditions can be based on attribute values such as priority, variable values such as number of declines, entity type or expressions, such as "NQ Process1.Queue" (Figure 5.6).

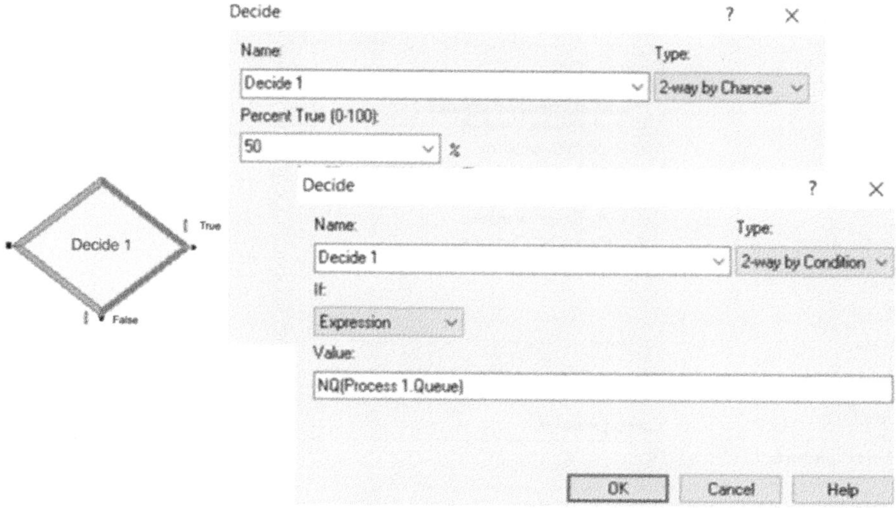

Figure 5.6. *"Decide" module*

This module has two or more output points depending on the specified type chosen by the user. The choice can be either "two-way by chance" or "two-way by condition" or even "N-way by chance" and "N-way by condition".

To create an expression when using a condition-based "Decide" module, the user must place the mouse pointer in the "Value" area and click on the right button. A new list then appears where the user must click on the "Build Expression" task to bring up the "Expression Builder" window (Figure 5.7). This tool includes several libraries of variables, statistical laws or mathematical functions that can be used when constructing the expression (Arena 2018).

5.3.1.6. *"Batch" module*

The "Batch" module serves to organize unit entities into batches within the simulation model (Arena 2018). The entities batches can be formed either as permanent or temporary (Figure 5.8).

It is important to highlight that the "Dispose" module does not support temporary batches.

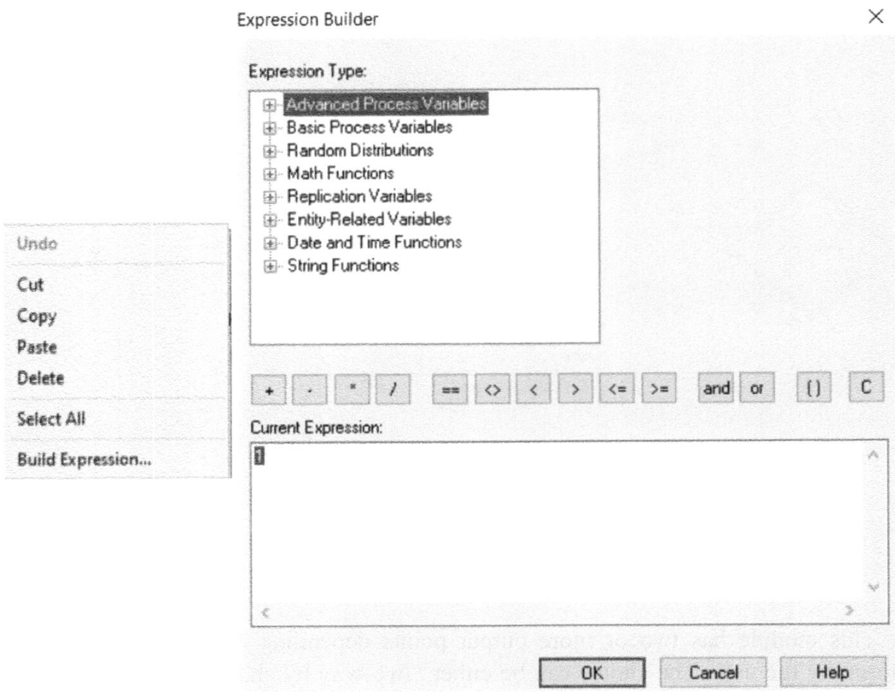

Figure 5.7. *Build expression tool*

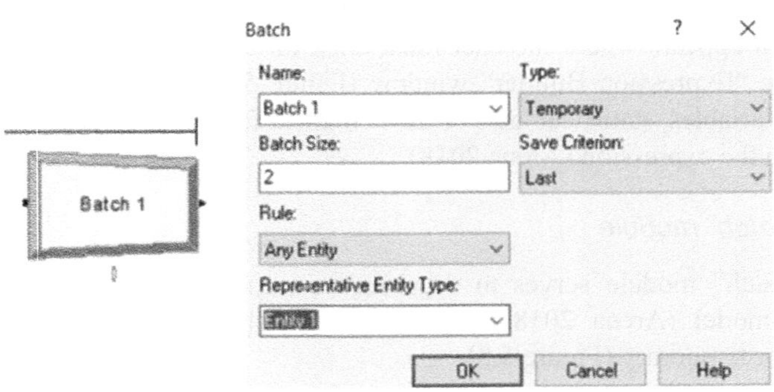

Figure 5.8. *"Batch" module*

Entities can be grouped in various arrangements, either arbitrarily or based on specific attributes. The entities are then held within a dedicated queue associated with the module, until the predefined threshold of entities is reached. Upon reaching this threshold, a new representative entity is generated and exits the module.

5.3.1.7. *"Separate" module*

The "Separate" module offers flexibility to either duplicate an incoming entity into several entities using the "Duplicate Original" function or split a pre-existing batched entities using the "Split Existing Batch" function (Arena 2018). Users can define rules governing the allocation of costs and times to the duplicates, as well as specify attribute assignment rules for the member entities (Figure 5.9).

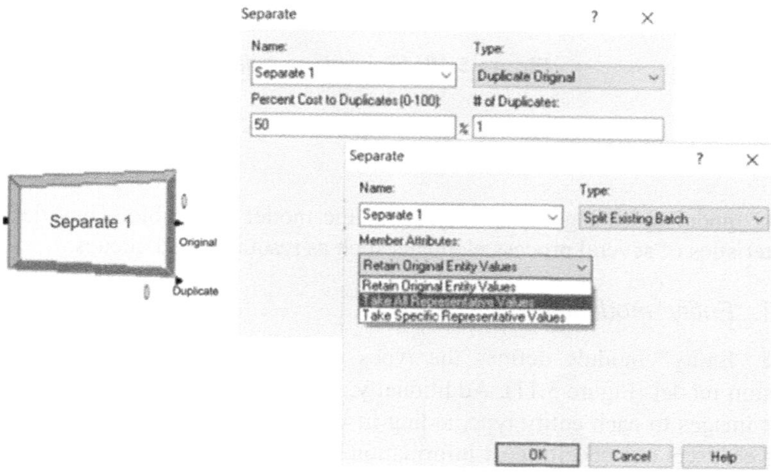

Figure 5.9. *"Separate" module*

To duplicate entities, the "Separate" module produces the designated number of copies, dispatching them alongside the departure of the original incoming entity. Conversely, the temporary representative entity that was formed during the batching is eliminated to divide the existing batch. The original entities that formed the batch are then recovered and extracted from the module in the same order in which they were initially grouped into batches. It is recommended to use the "Take All Representative Values" option so that entities retain the updated values of their attributes and variables assigned to them.

5.3.1.8. *"Record" module*

The "Record" module is used to collect statistics within the simulation model. It enables the collection of diverse observations, including time intervals between outputs and entity-specific statistics like time and cost. Moreover, it offers the flexibility to organize collected statistics into intervals. Additionally, serving as a counter, it systematically increments a statistical value with each passage of an entity, providing valuable data for analysis and decision-making (Arena 2018) (Figure 5.10).

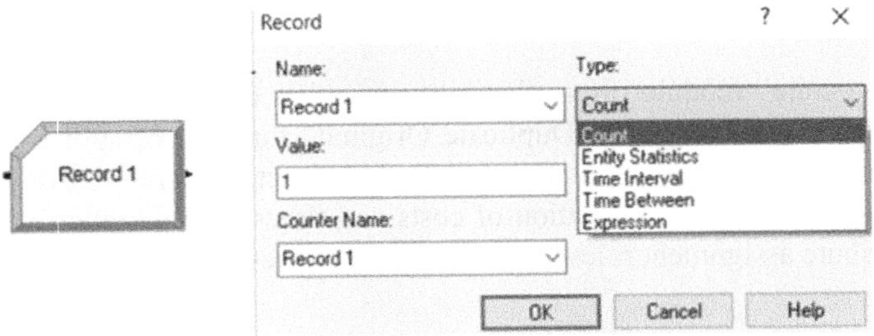

Figure 5.10. *"Record" module*

5.3.2. *Data modules*

Data modules are a group of objects in the model data table. They define the characteristics of several process elements such as resources and queues.

5.3.2.1. *"Entity" module*

The "Entity" module defines the types of the different entities used in the simulation model (Figure 5.11). Additionally, it provides the functionality to assign distinct images to each entity type, aiding in visual clarity during model animation. Moreover, users can specify cost information for individual entities, contributing to more comprehensive cost analysis within the simulation (Arena 2018).

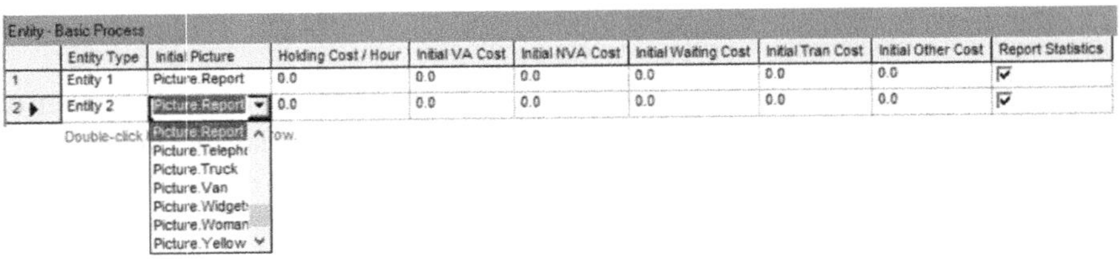

Figure 5.11. *"Entity" module*

5.3.2.2. *"Resource" module*

Within this data module, the groundwork for resource definition in the simulation system is laid out (Figure 5.12). Here, resources are delineated, with options for fixed capacities or scheduled operations. This module also accommodates the referencing of failures and resource states for further use with the "Advanced Process " and "Advanced Transfer" panels (Arena 2018).

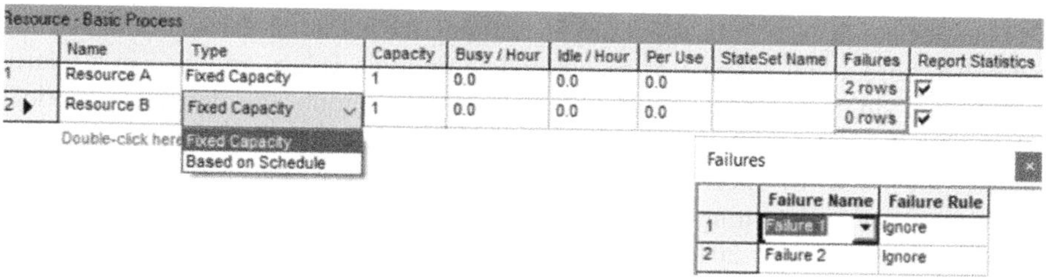

Figure 5.12. *"Resource" module*

5.3.2.3. *"Queue" module*

The "Queue" data module serves to define queues within the simulation model (Figure 5.13). It provides the flexibility to adjust the scheduling rule for each queue, with the default setting being "First In First Out". It can then easily be modified to "First In Last Out". Queue control rules can also be based on attributes allowing for dynamic prioritization. This involves choosing the type of condition in relation to this attribute, i.e. its lowest value, "Lowest Attribute Value", or the largest, "Highest Attribute Value". For example, the "Short Processing Time" rule involves choosing the "Lowest Attribute Value" mode for the "manufacturing time" attribute of the entities waiting in the queue (Arena 2018).

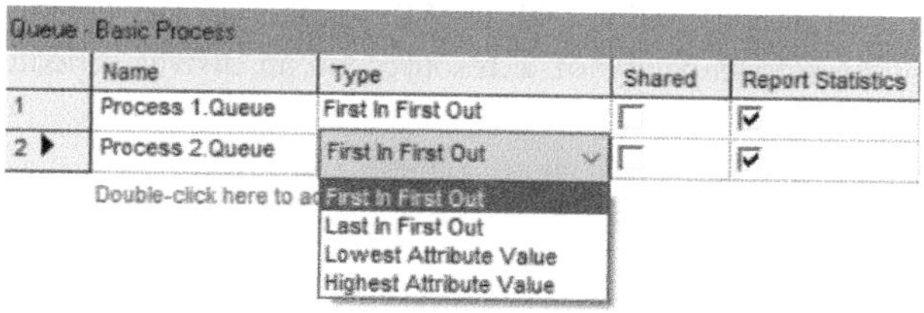

Figure 5.13. *"Queue" module*

There is an additional field that allows sharing the queue when it must be used in several places in the simulation model.

5.3.2.4. *"Variable" module*

The main purpose of the "Variable" data module is to define the characteristics of the variables, including their dimensions and specific values (Figure 5.14). These variables play a pivotal role in various modules, such as the "Decide" module. Additionally, the variables can be reassigned new values with the "Assign" module as needed and used in any expression.

One notable feature is the flexibility in defining the nature of these variables. They can either store a single value or accommodate multiple values arranged in a matrix format. For single-valued variables, the user simply inputs the desired numerical value directly in the "Initial Values" cell. Conversely, for multiple-value variables, the user must start by specifying the values of the "Rows" and "Columns" making up the matrix, then inputting the values of the variables in the corresponding cells of "Initial values" (Arena 2018). This versatility empowers users to model a wide range of scenarios and expressions within the system .

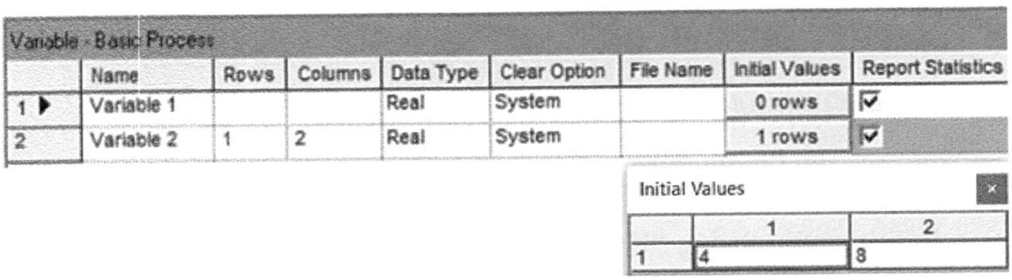

Figure 5.14. *"Variable" module*

5.3.2.5. *"Schedule" module*

This "Schedule" data module is used to define an operating or an arrival schedule. It hence has to be used in conjunction with the "Resource" or with the "Create" module to define the processing capacity of a resource or an arrival schedule for entities (Arena 2018) (Figure 5.15).

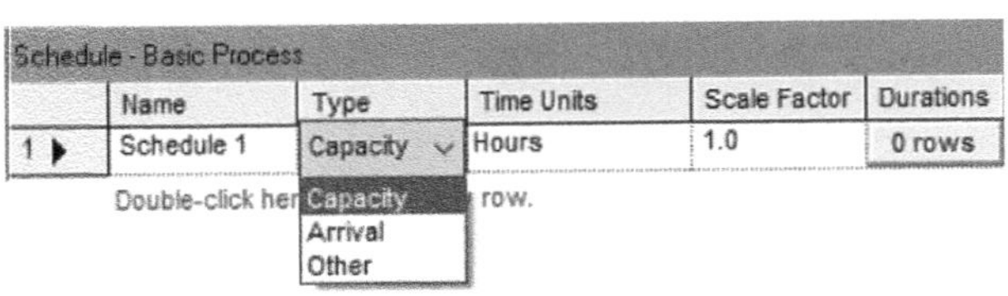

Figure 5.15. *"Schedule" module*

5.4. The "advanced process" panel

5.4.1. *Flowchart modules*

5.4.1.1. *"Delay" module*

This module can be used to postpone the progression of any entity by a specified duration. Upon an entity's arrival at this module, it undergoes a specified expression of the waiting time. Subsequently, the entity is retained by this module until the prescribed time period expires. This duration can be assigned to the entity as value added, transfer, wait or according to other specification. Moreover, associated costs can be computed if assigned (Figure 5.16).

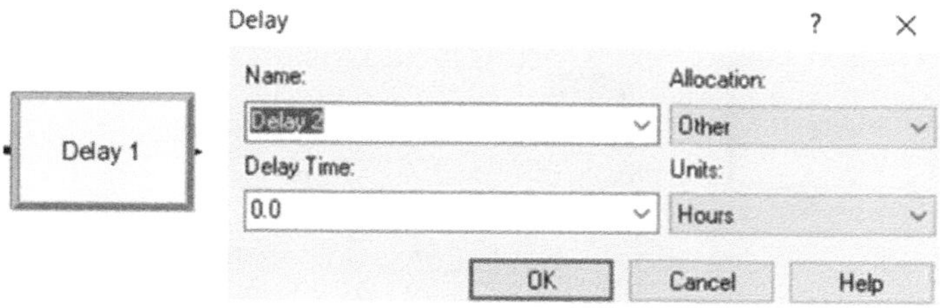

Figure 5.16. *"Delay" module*

5.4.1.2. *"Hold" module*

This module keeps any incoming entity in a queue, either to wait for a signal, to wait for a specified condition to become true, or indefinitely (Figure 5.17).

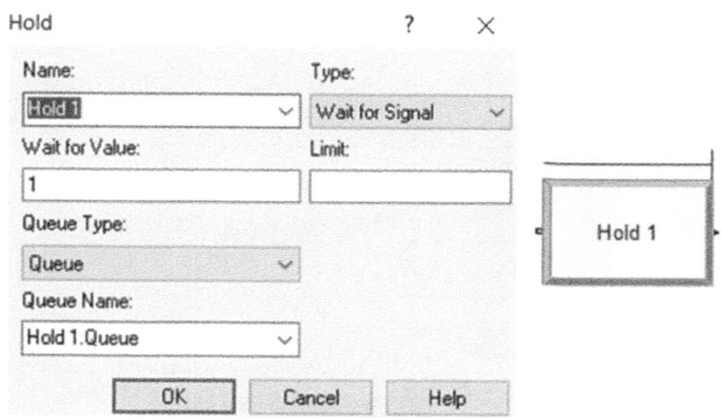

Figure 5.17. *"Hold" module*

When an entity awaits a signal to proceed, it engages the "Signal" module, facilitating its progression to the next module. Conversely, if it anticipates a specific condition to be true, it persists within the module until the condition is fulfilled. In the case where the entity is put on indefinite hold, the "Remove" module intervenes to allow the entity to continue processing (Arena 2018).

5.4.1.3. *"Pickup" module*

Whenever an entity enters the "Pickup" module, it systematically extracts a defined number of consecutive entities from a specified queue, starting from a determined rank in that queue (Arena 2018). The extracted entities are subsequently added to the original entities that initiated this module (Figure 5.18).

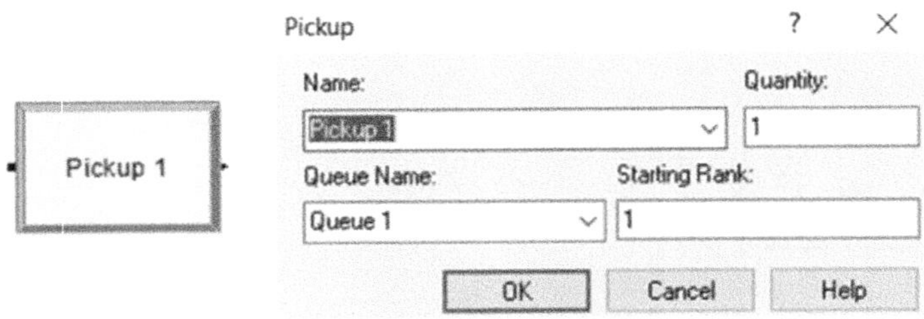

Figure 5.18. *"Pickup" module*

5.4.1.4. *"ReadWrite" module*

The "ReadWrite" module serves the dual purpose of reading data values from an input file and assigning these data values to a list of variables or attributes in the model (Arena 2018). Moreover, it facilitates the transfer of simulation model data to an external file (Figure 5.19).

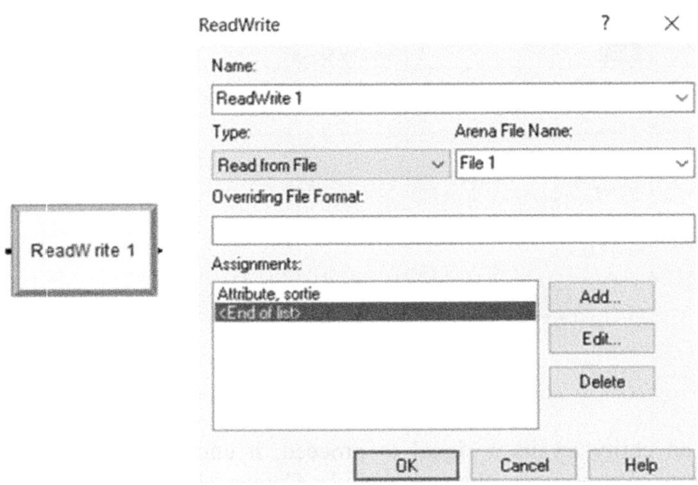

Figure 5.19. *"ReadWrite" module*

5.4.1.5. *"Remove" module*

The "Remove" module operates by extracting a singular entity from a predefined position within a queue and redirecting it to a designated module (Figure 5.20). Upon arrival at a "Remove" module, the specified entity is promptly taken out from the queue and directed towards its designated destination module. The term "Rank of Entity" signifies the precise location of the entity within the queue. The original entity that initiated the removal process is subsequently routed to another module corresponding to the destination of the extracted entity (Arena 2018).

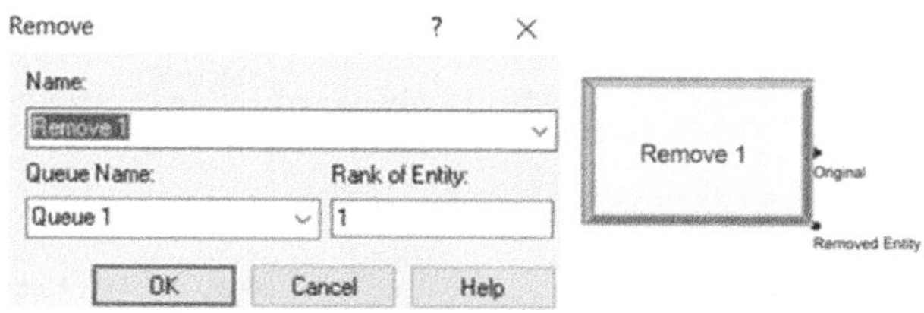

Figure 5.20. *"Remove" module*

5.4.1.6. *"Signal" module*

The "Signal" module dispatches a designated value as a signal to every "Hold" module within the model configured to "Wait for Signal". These modules must therefore release the number of entities specified in the "Limit" (Figure 5.21).

Upon an entity's arrival at this module, the signal undergoes evaluation, prompting the transmission of the corresponding signal code. Subsequently, entities in "Hold" modules waiting for the same signal are released from their queues. The entity responsible for transmitting the signal proceeds with its processing until its eventual removal from the system (Arena 2018).

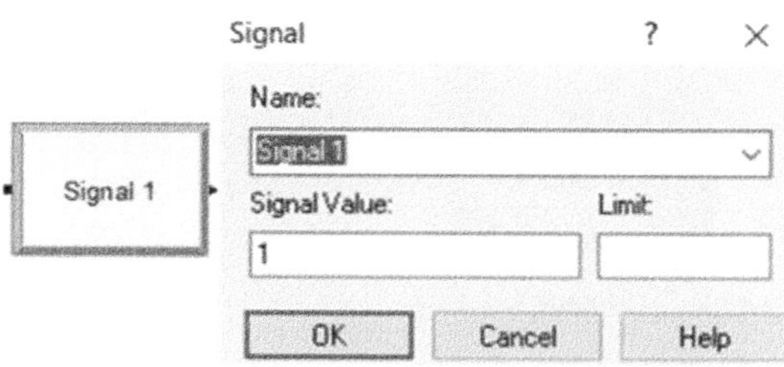

Figure 5.21. *"Signal" module*

5.4.1.7. *"Store" module*

Upon an entity's arrival at a "Store" module, a designated storage unit undergoes incrementation. This entity immediately moves to the next module within the model. Storages are useful for displaying the animation of an entity while it undergoes processing across other modules (Arena 2018). Additionally, they facilitate the maintenance of statistical records concerning the quantity of entities stored (Figure 5.22).

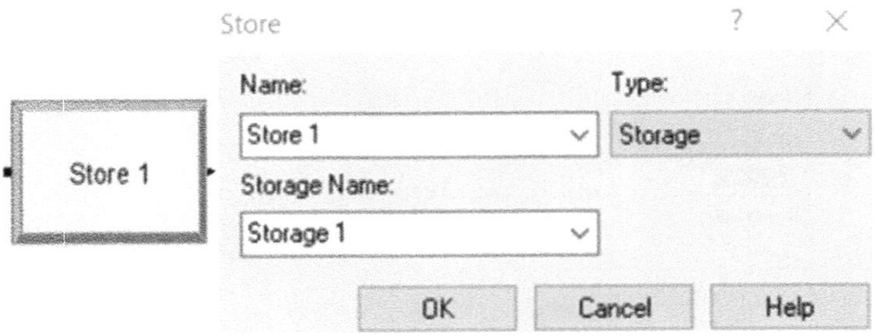

Figure 5.22. *"Store" module*

5.4.1.8. *"Unstore" module*

The "Unstore" module deletes an entity from storage. Upon an entity's arrival at this module, the designated storage experiences a decreased decrement and the entity immediately moves to the subsequent module within the model (Figure 5.23).

Entities entering this module are expected to have been previously stored in the specified storage. Otherwise, a warning will be triggered. Additionally, before disposal, entities must be extracted from all storages in which they were stored (Arena 2018).

Figure 5.23. *"Unstore" module*

5.4.1.9. *"Search" module*

The "Search" module scans a queue, seeking an entity that meets the prescribed search criteria (Figure 5.24).

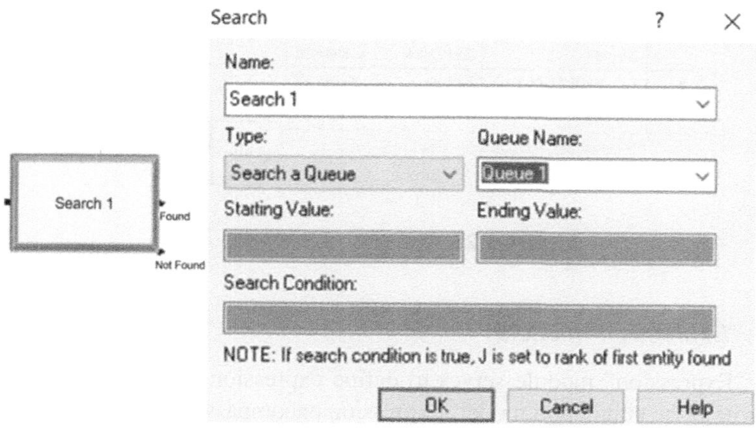

Figure 5.24. *"Search" module*

During the search operation, a global system variable labeled as "J" is initiated to the rank of the first entity that fulfills the specified search criteria. If no entity meets the criteria, "J" is set to 0. Upon an entity's arrival at a search module, the index "J" is set to the starting index and the search condition is then assessed. If the search condition is satisfied, the search concludes, and the updated value of "J" is retained. Otherwise, the value of "J" is increased or decreased and the condition is checked again. This process is repeated until the search condition is satisfied or the final value is reached. If the condition is not fulfilled or there are no entities in the queue or group, "J" retains a value of 0 (Arena 2018).

5.4.2. *Data modules*

5.4.2.1. *"Advanced Set" module*

The "Advanced Set" module serves to define various sets, including queue sets, storage sets and other sets, along with their respective members. These sets are designed to encapsulate similar elements under a unified name and set index, enabling streamlined referencing. The elements that make up the set are collectively called the members of the set (Figure 5.25).

Queue sets can be specifically designated within modules of the type "Seize", while storage sets find utility in modules like "Store" and "Unstore". Additionally, other sets might encompass the collection of attributes or other miscellaneous elements (Arena 2018).

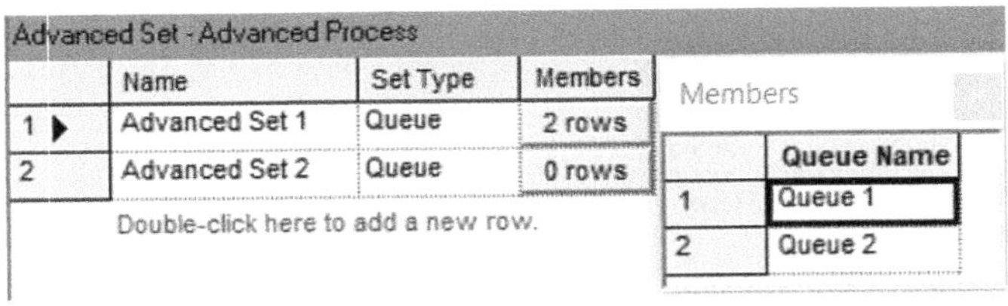

Figure 5.25. *"Advanced Set" module*

5.4.2.2. *"Expression" module*

The "Expression" module serves to define expressions and their corresponding values within the simulation model. Expression encompasses a wide range of logical operations supported by the software, including real values, distributions, variables and combinations of these (Arena 2018). These expressions can manifest as single elements or extend to one- or two-dimensional arrays (Figure 5.26).

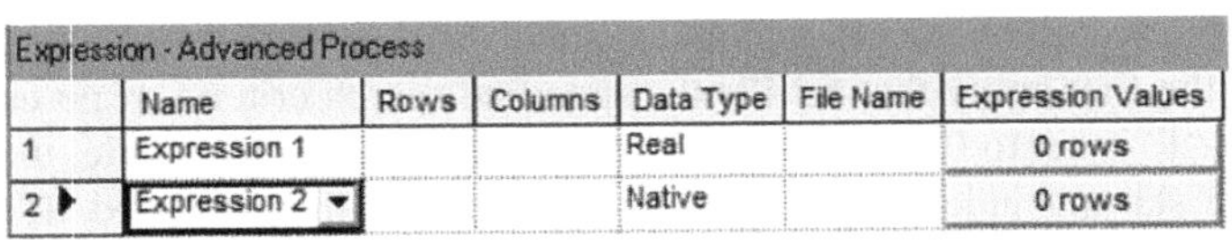

Figure 5.26. *"Expression" module*

5.4.2.3. *"Failure" module*

The "Failure" module is specifically developed to operate in conjunction with resources. Upon occurrence of a failure event, the entire resource is designated as failed (Arena 2018). Failure care is intended to be utilized with either single-capacity resources or with multi-capacity resources where all individual resource units fail concurrently (Figure 5.27).

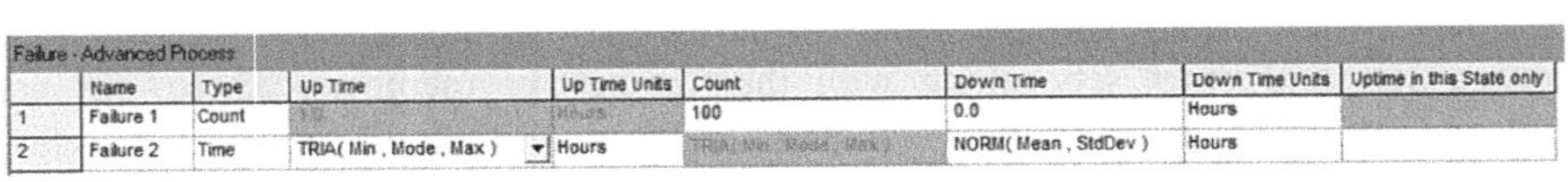

Figure 5.27. *"Failure" module*

5.4.2.4. *"File" module*

The "File" module serves to facilitate access to external files through interaction with the "ReadWrite" module (Arena 2018). It is responsible for specifying the name of the system file and establishing the access type and operational attributes associated with the file (Figure 5.28).

Figure 5.28. *"File" module*

5.4.2.5. *"Statistics" module*

The "Statistics" module serves to define supplementary statistics gathered throughout the simulation and in specifying the output data files (Arena 2018) (Figure 5.29).

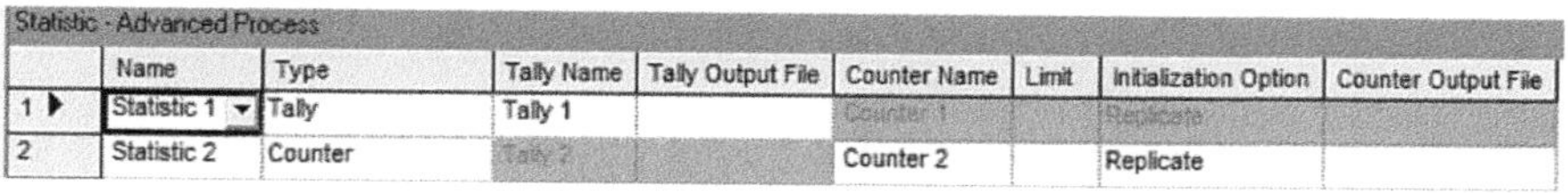

Figure 5.29. *"Statistics" module*

5.5. The "advanced transfer" panel

5.5.1. *General flowchart modules*

5.5.1.1. *"Station" module*

The "Station" module delineates a station representing either a physical or logical location where processing occurs (Arena 2018). In cases where the "Station" module specifies a collection of stations, it essentially outlines multiple treatment locations (Figure 5.30).

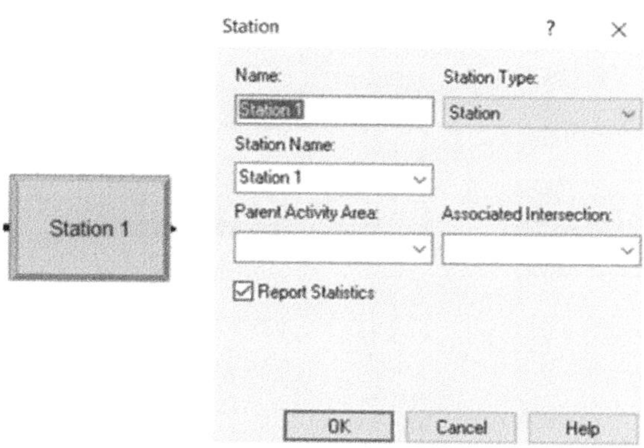

Figure 5.30. *"Station" module*

5.5.1.2. *"Route" module*

The "Route" module facilitates the movement of an entity to a designated station or the subsequent station in the sequence defined for the entity (Figure 5.31). Optionally, a delay for transfer to the next station can be configured. Upon entering the "Route" module, the "Entity.Station" attribute is configured as the destination station. Subsequently, the entity is directed to the destination station, adhering to the specified transit time. If the station destination is specified as "By Sequence", the next station is determined based on the entity's sequence and step within the set, as defined respectively by the special attributes "Entity Sequence" and "Entity.Jobstep" (Arena 2018).

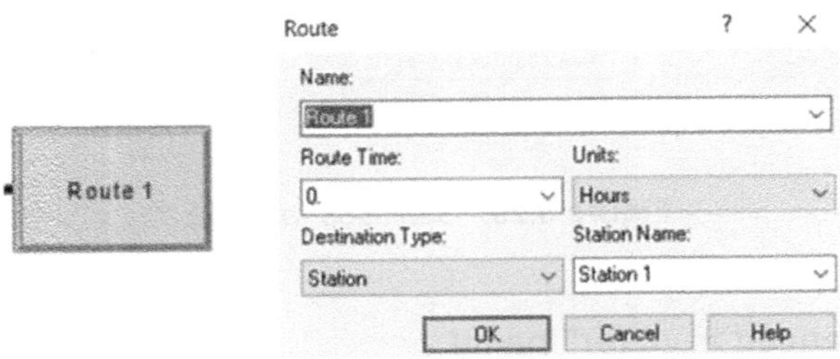

Figure 5.31. *"Route" module*

5.5.1.3. *"PickStation" module*

The "PickStation" module allows an entity to choose a specific station from several options provided (Figure 5.32).

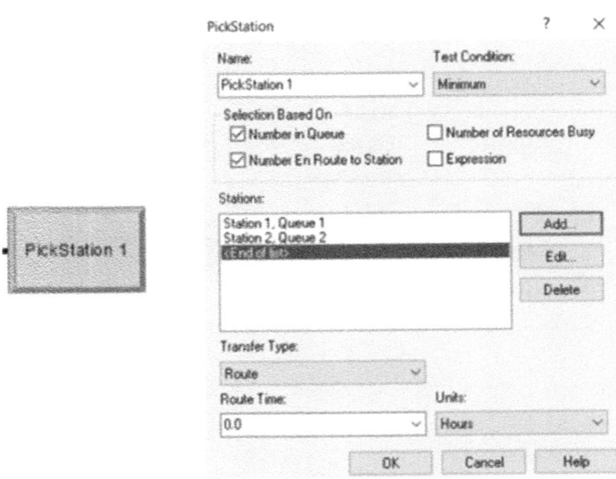

Figure 5.32. *"PickStation" module*

This module makes a choice from the group of stations based on the selection logic defined within. Subsequently, the entity can be directed, transported or connected to the chosen station. If connection is the chosen method, the selected station is attributed to an entity attribute. Station selection is based on the minimum or maximum value of a variety of predefined variables, such as the number of entities in queue, the number of entities en route to the station or the number of busy resources. Moreover, the "Build Expression" tool can be utilized to construct an expression on the basis of which the "PickStation" module chooses the station (Arena 2018).

5.5.2. *Flow conveyor modules*

5.5.2.1. *"Convey" module*

The "Convey" module efficiently transfers an entity along a conveyor system from its current position to a specified destination station. The duration of the transport between stations is influenced by both the conveyors and the distance between stations, parameters that must be defined respectively within the "Convey" and "Segment" modules (Figure 5.33).

Upon entry into the "Convey" module, its "Entity.Station" attribute is adjusted to match the destination station. Subsequently, the entity is transported to the destination station. If the destination type is set as "By sequence", the subsequent station is determined based on the sequence and the entity's position within the sequence (Arena 2018).

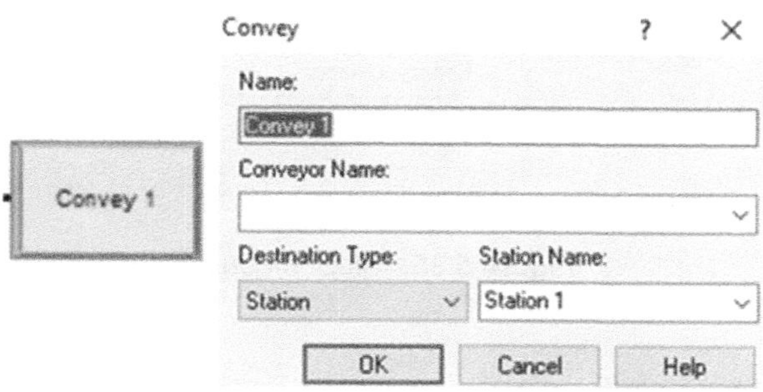

Figure 5.33. *"Convey" module*

5.5.2.2. *"Start" module*

The "Start" module initiates the transition of a conveyor from inactive to active (Arena 2018). This activation can occur after the conveyor has been previously deactivated by the "Stop" module or when it was initially set to an inactive state at the beginning of the simulation. Additionally, upon activation, there is an option to permanently adjust the conveyor speed (Figure 5.34).

Figure 5.34. *"Start" module*

5.5.2.3. *"Stop" module*

The "Stop" module determines the operational status of a conveyor as inactive. This status change can occur either after activation from the "Start" module or when initially set as active at the start of the simulation. Upon entry into the "Stop" module by an entity, regardless of the conveyor type or the number of entities present, the conveyor halts immediately (Figure 5.35).

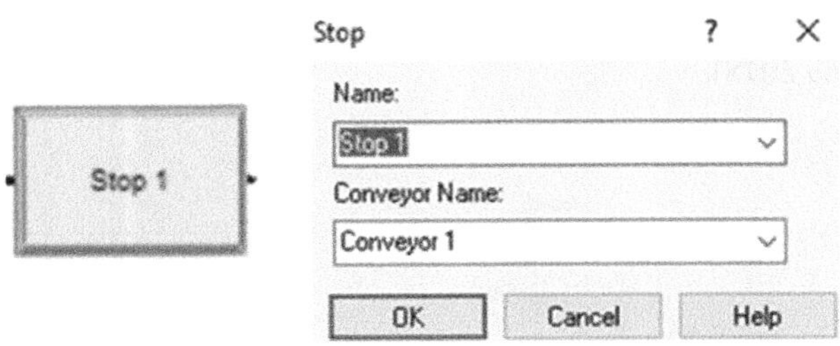

Figure 5.35. *"Stop" module*

5.5.2.4. *"Access" module*

The "Access" module allocates one or multiple cells of a conveyor to an entity, enabling its transfer from one station to another. Once the entity has secured control over the conveyor cells, it becomes eligible for transportation to the subsequent station (Arena 2018). Upon arrival at an "Access" module, an entity awaits until the appropriate number of contiguous conveyor cells are vacant and properly aligned with the entity's station location (Figure 5.36).

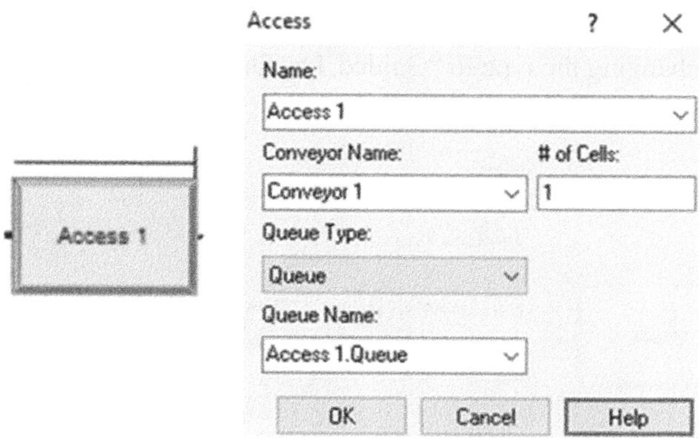

Figure 5.36. *"Access" module*

5.5.2.5. *"Exit" module*

The "Exit" module discharges the entity's cells on the specified conveyor (Arena 2018). Should another entity be queued at the same station, it promptly gains access to the conveyor upon the discharge of cells (Figure 5.37).

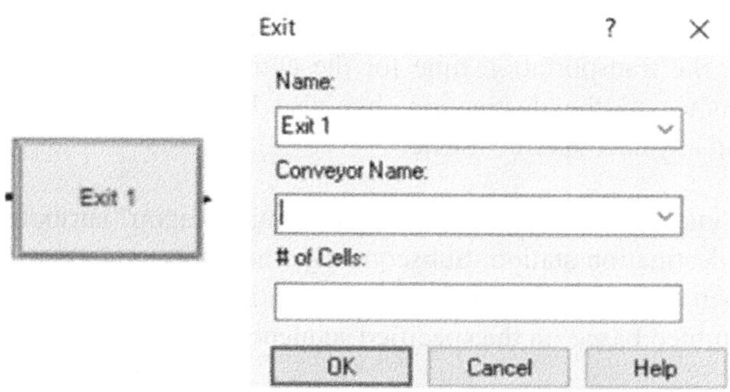

Figure 5.37. *"Exit" module*

5.5.3. *Flow transport modules*

5.5.3.1. *"Transport" module*

The "Transport" module transfers the control entity to its designated destination station. Following the required delay for transport, the entity reappears in the model at the "Station" module. Free-path transporters consistently move to the same destination station as the feature. However, guided transporters offer flexibility, allowing relocation to a network location distinct from the entity's station destination by changing the type to "Guided TranDestination" (Figure 5.38).

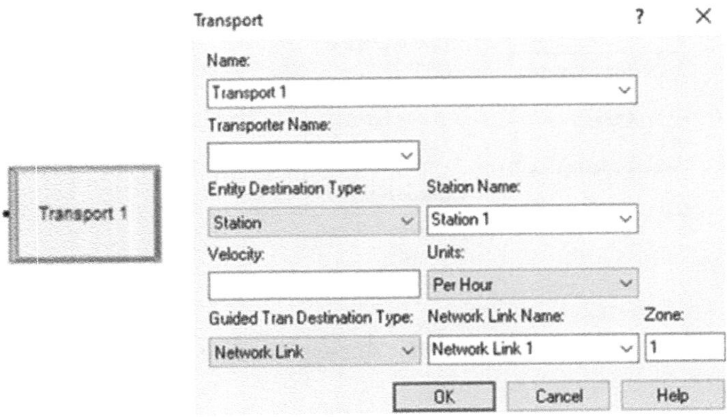

Figure 5.38. *"Transport" module*

When a free-path transporter is used, the duration for transporting the entity from one station to another is based on the speed of the transporter and the distance between stations as specified in the "Distance" module. Conversely, if the guided carrier is used, the transportation time for the entity is influenced not only by the speed and distance to the destination, but also by factors such as acceleration, deceleration and any encountered traffic.

Upon entry into the "Transport" module, its "Entity.Station" attribute is configured to the entity's destination station. Subsequently, the entity undergoes transportation to the destination station. If the destination station is set as "By Sequence", the next station is determined based on the specified sequence (Arena 2018).

5.5.3.2. *"Request" module*

The "Request" module allocates a transport unit to any requesting entity. It moves the allocated unit to the entity's location (Figure 5.39).

The selection of the transport unit can be dynamically determined based on a rule. Upon arrival at the request module, the entity is promptly assigned to a carrier, if it is available. Subsequently, the entity waits within the request module until the transport unit reaches its location. Upon completion of the journey, the entity exits the request module (Arena 2018).

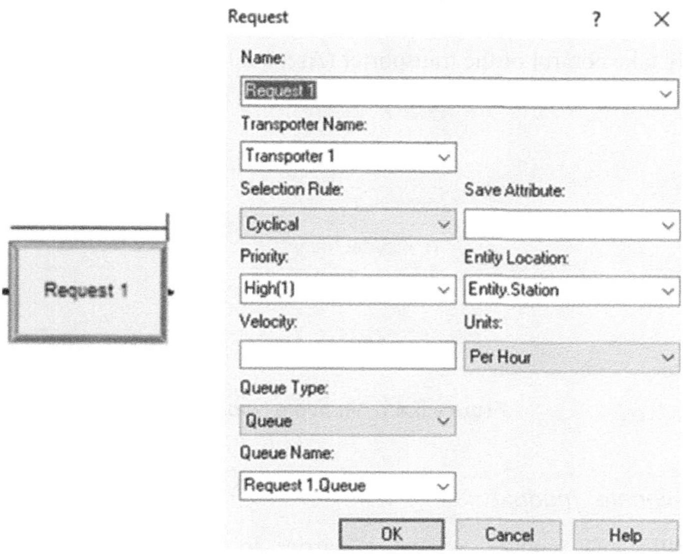

Figure 5.39. *"Request" module*

5.5.3.3. *"Free" module*

The "Free" module releases the transport unit that was allocated to the entity by a "Request" module (Figure 5.40). If no entity requests this released transport unit, it will remain waiting at the station location of the "Free" module, unless otherwise specified in the "Transport" module that dispatched this transport unit (Arena 2018).

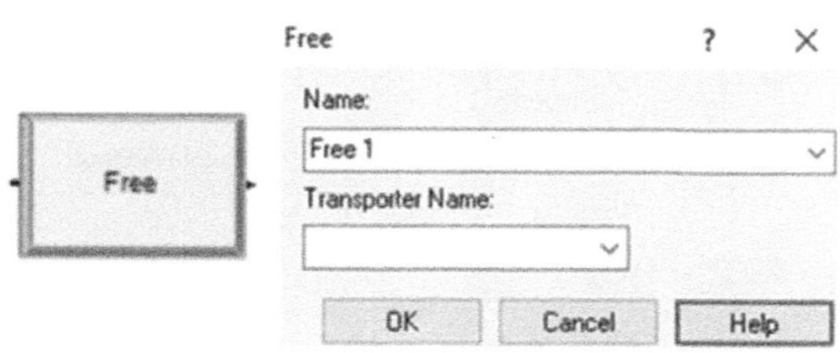

Figure 5.40. *"Free" module*

5.5.3.4. *"Activate" module*

The "Activate" module enhances the capacity of a previously halted or initially inactive transporter, as per the specifications in the "Transport" module (Figure 5.41). Once activated, the transport unit remains at the location of the station where it was stopped until either relocated or requested by an entity. In instances where one or more entities await a transporter at the time of activation, the entity will immediately take control of the transporter (Arena 2018).

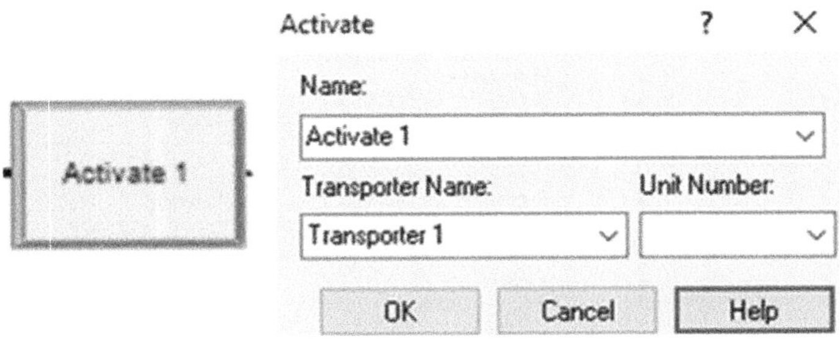

Figure 5.41. *"Activate" module*

5.5.3.5. *"Allocate" module*

The "Allocate" module assigns a carrier to an entity without physically relocating it to the entity's station location (Arena 2018). Subsequently, the entity gains control over the transporter to move it to a specified destination or halt the transporter in case of malfunction or failure (Figure 5.42). The assignment of a specific transport unit to the entity is possible, or a rule-based selection process can determine which carrier will be allocated to the entity.

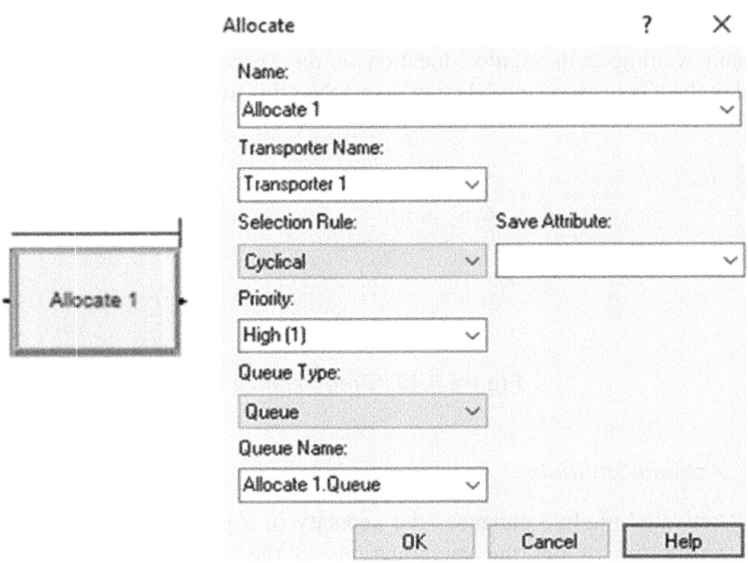

Figure 5.42. *"Allocate" module*

5.5.3.6. *"Halt" module*

The "Halt" module transitions the status of a transport unit from active to inactive. If the carrier is engaged when an entity enters the "Halt" module, its status is marked as occupied and inactive until the controlling entity releases the unit (Arena 2018). Once a transport unit has been halted, no entity can assume control of the transport until it is reactivated (Figure 5.43).

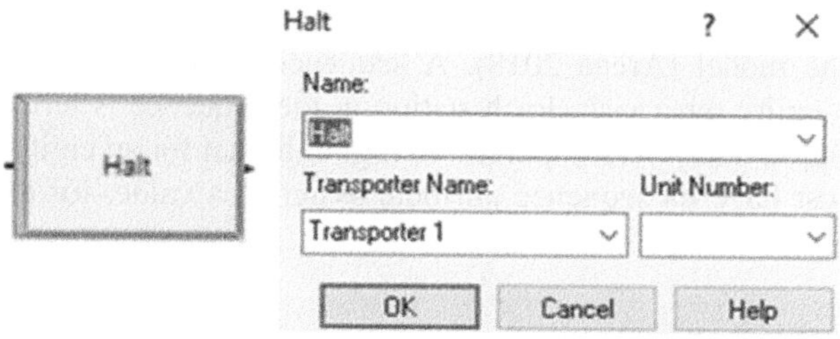

Figure 5.43. *"Halt" module*

5.5.3.7. *"Move" module*

The "Move" module relocates a transporter from one location to another without physically moving the controlling entity to the destination station. The controlling entity stays at the current location within the module until the transporter reaches its destination (Arena 2018). Once the transporter arrives, the entity gains the flexibility to proceed to another module or task within the model (Figure 5.44).

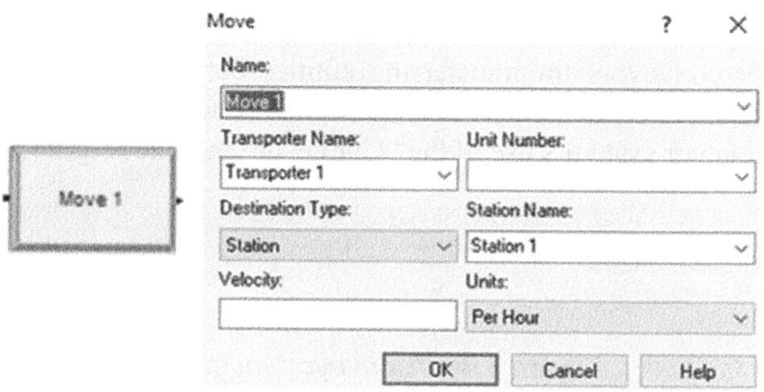

Figure 5.44. *"Move" module*

If the carrier being moved is a free-path carrier, the duration of its movement from one station to another is determined by the carrier's speed and the distance between stations specified in the "Distance" module. However, if the carrier is a guided one, its movement time is influenced not only by speed and distance to the destination but also by factors such as acceleration, deceleration and potential traffic encountered along the route.

5.5.4. *Data modules*

5.5.4.1. *"Sequence" module*

The "Sequence" module is employed to establish the processing sequence of an entity within the model (Arena 2018). A sequence comprises a structured list of stations that an entity must visit. Each station in the sequence is termed a "step" or "work step" (Figure 5.45). It is important to highlight that for an entity to adhere to a sequence, it must have its sequence attribute assigned a value, for example, in the "Assign" module.

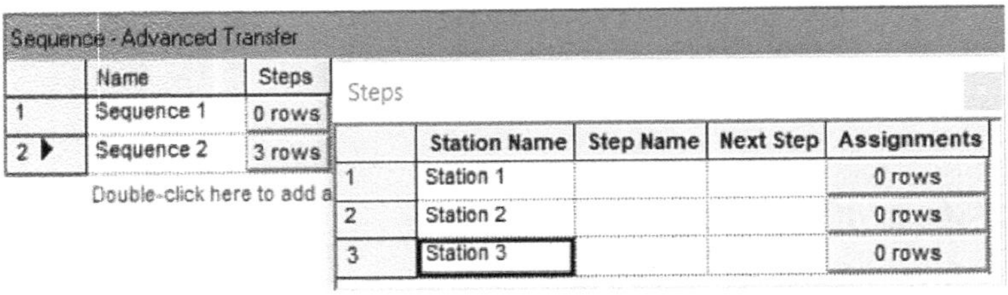

Figure 5.45. *"Sequence" module*

5.5.4.2. *"Conveyor" module*

The "Conveyor" module enables the configuration of both accumulator or non-accumulator conveyors for transferring entities between stations (Arena 2018). When employing a conveyor for transferring entities between modules, it is imperative to define the conveyor system's use of the "Conveyor" module (Figure 5.46).

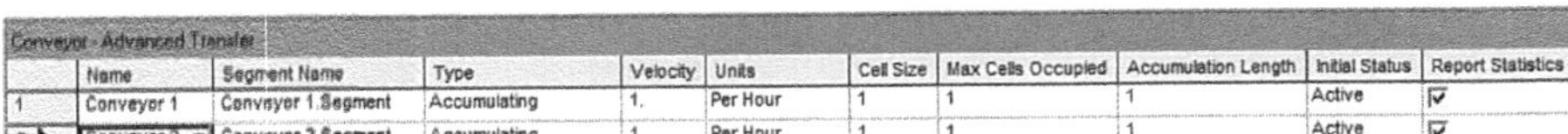

Figure 5.46. *"Conveyor" module*

5.5.4.3. *"Segment" module*

The "Segment" module precisely delineates the distance between two stations within a network of paths for a conveyor system (Arena 2018). By specifying the start station, end station and path, an appropriate set of paths is generated, effectively defining the trajectory of the conveyor (Figure 5.47).

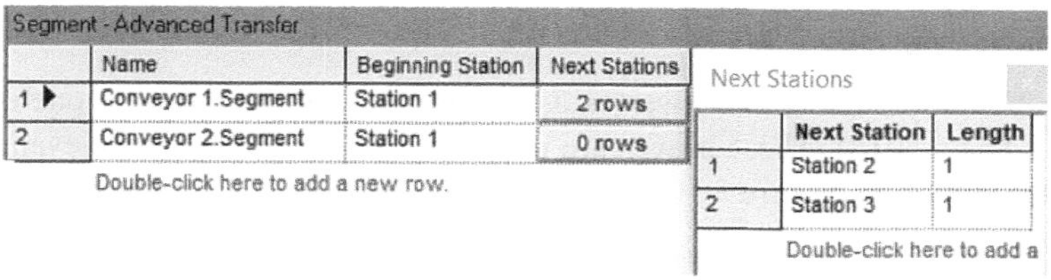

Figure 5.47. *"Segment" module*

5.5.4.4. *"Transport" module*

The "Transport" module allows the definition of transport mechanisms capable of execution either freely or with a guided trajectory for relocation of entities between different locations. Free-path carriers autonomously move between stations and are not influenced by the traffic of other carriers. The "Distance" module specifies the distances between stations. Conversely, guided transporters are limited to predetermined trajectories such as tracks or rails and may experience delays due to congestion caused by other vehicles (Figure 5.48) (Arena 2018).

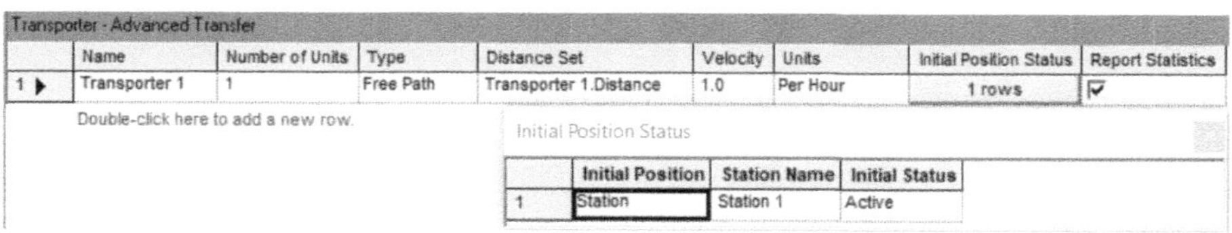

Figure 5.48. *"Transport" module*

5.5.4.5. *"Distance" module*

The "Distance" module plays a role in defining the distances covered between every pair of stations (Arena 2018). It features the distance name along with a detailed list of station pairs and their corresponding individual distances (Figure 5.49).

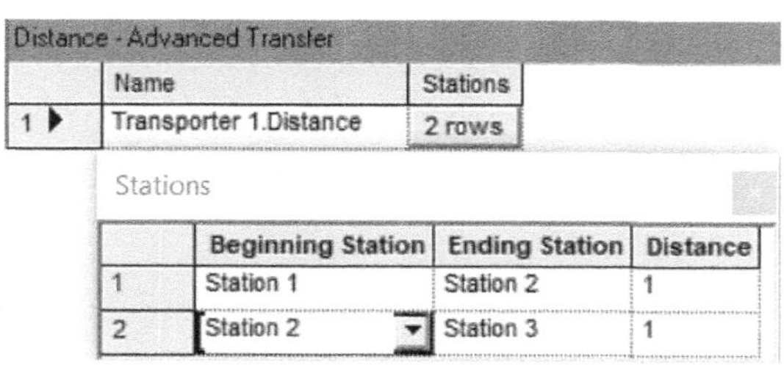

Figure 5.49. *"Distance" module*

5.5.4.6. *"Activity Area" module definition*

The "Activity Area" module serves as a space that can be linked to one or more stations (Arena 2018). These spaces are used for collecting statistics relevant to the associated station(s) (Figure 5.50).

Activity Area - Advanced Transfer

	Name	Description	Organization Level	Parent Activity Area	Report Statistics
1	Activity Area 1		100		☑
2 ▶	Activity Area 2		100		☑

Figure 5.50. *"Activity Area" module*

6

Simulation Modeling

6.1. Introduction to using Arena software

6.1.1. *Description and data*

This first manipulation serves as an introduction to the use of the Arena simulation software. It involves building a first simple simulation model and comparing its results with those determined manually.

A manufacturing workshop produces a single type of part using two machines M_1 and M_2 placed in series. Parts arrive in the production workshop individually, with a consistent inter-arrival time of 16 minutes. It is worth noting that the first part appears at the entrance to machine M_1 at $t = 0$. Once the part is manufactured by the M_1 machine, it will be transferred to the M_2 machine. The parts manufacturing times on the machines are respectively 15 and 20 minutes. Transfer times between the two machines are considered negligible.

(1) Manual simulation: for a manufacturing order of 100 parts, calculate the following performance indexes on the basis of a manual simulation by running the model without using the simulation software:

– the total time to manufacture the entire order;

– the maximum number of parts waiting in each machine queue.

(2) Simulation with Arena software: using Arena software, develop the simulation model relating to this problem using the appropriate modules with a model stopping condition corresponding to the manufacture of 100 parts. Subsequently, simulate the model and analyze the results.

(3) Compare the simulation results with Arena with those of the manual simulation.

6.1.2. *Manual simulation*

The time required for manufacturing the order: the first part leaves the workshop at t = 35 minutes, the second part leaves at t = 55 = 35 + 20 minutes, the third part leaves at t = 75 = 55 + 20 minutes, the fourth one leaves at t = 95 = 75 + 20 minutes and so on. It is notable that if a part is manufactured by the two machines at time T, the subsequent part's manufacturing will be completed at t = T + 20 minutes. As a result, the exit time of part "i" from the manufacturing workshop is: $T_i = T_{i-1} + 20 = T_{i-2} + 20 \times 2 = \ldots = T_1 + 20 \times (i-1)$. The time required to manufacture the entire order of 100 parts will thus be $T_{100} = T_1 + 20 \times (100-1) = 35 + 20 \times 99 = 2{,}015$ minutes (Figure 6.1).

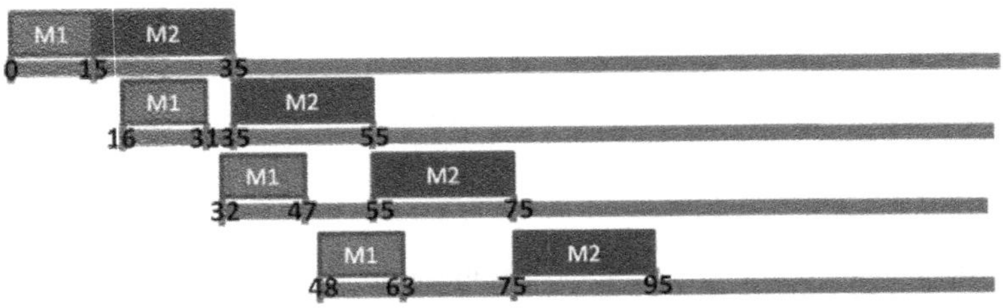

Figure 6.1. *Manual simulation graph*

For the maximum number of parts waiting in front of each machine, we can note that there is no waiting in front of the first machine, M_1. This is due to the fact that the inter-arrival time of parts, equal to 16 minutes, is greater than the manufacturing time on this machine which is 15 minutes.

In the case of the second machine, M_2, the waiting time for the second part is equal to the time necessary for the manufacture of the first part by the two machines minus the time necessary for the manufacture of this second part by the machine M_1 from its arrival at t = 16 minutes. So, the waiting time for the second part is $WT_{M2}(P_2) = (15 + 20) - (16 + 15) = 4$ minutes. The waiting time for the third part is equal to the time necessary for the manufacture of the second part by the two machines minus the time necessary for the manufacture of this third part by the machine M_1 from its arrival at t = 32 minutes. So, $WT_{M2}(P_3) = (35 + 20) - (32 + 15) = 8 = 2 \times 4$ minutes.

The waiting time for the fourth part is equal to the time necessary for the manufacture of the third part by the two machines minus the time necessary for the manufacture of this fourth part by the machine M_1 from its arrival at t = 48 minutes. So, $WT_{M2}(P_4) = (55 + 20) - (48 + 15) = 12 = 3 \times 4$ minutes. As a result, the waiting time of a part "i" in front of the machine M_2 is equal to $WT_{M2}(P_i) = (i - 1) \times 4$ minutes. In addition, the maximum waiting time is reached for the last part with

WT_{M2} (P_{100}) = 99 x 4 = 396 minutes. The maximum number of parts waiting is obtained by dividing this maximum waiting time by the manufacturing time for a part by the machine M_2: $\frac{396}{20} \approx 20$ parts.

6.1.3. *Simulation with Arena software*

6.1.3.1. *Simulation model*

The simulation model is composed by a "Create" module, named "parts arrivals", followed by two "Process" modules placed in series and named respectively "machine M1" and "machine M2". The model concludes with a "Dispose" module called "parts exit" (Figure 6.2).

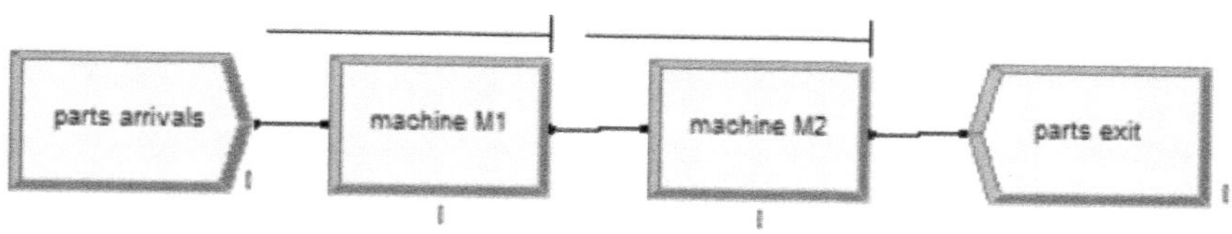

Figure 6.2. *Simulation model*

6.1.3.2. *Configuration of the "Create" module*

Figure 6.3 shows the configuration of the "Create" module.

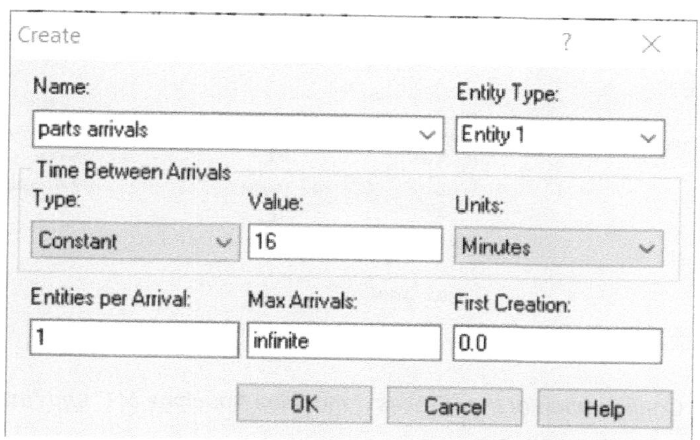

Figure 6.3. *Configuration of the "Create" module "parts arrival"*

Based on the model description and data, the type of inter-arrival time "Time Between Arrivals" is constant and must be set to 16 in the "Value" box. It is important to specify that the time units are in minutes in the "Units" box. To comply with the condition of presentation of the first part at t = 0, we simply specify that the first creation takes place at t = 0.0 in the "First Creation" box.

6.1.3.3. *Configuration of "Process" modules*

When modeling the two machines using the "Process" modules, two different resources must be used for the two machines, "Resource, M1, 1" and "Resource, M2, 1"; otherwise, it will involve the use of the same resource in two different locations alternatively (Figure 6.4).

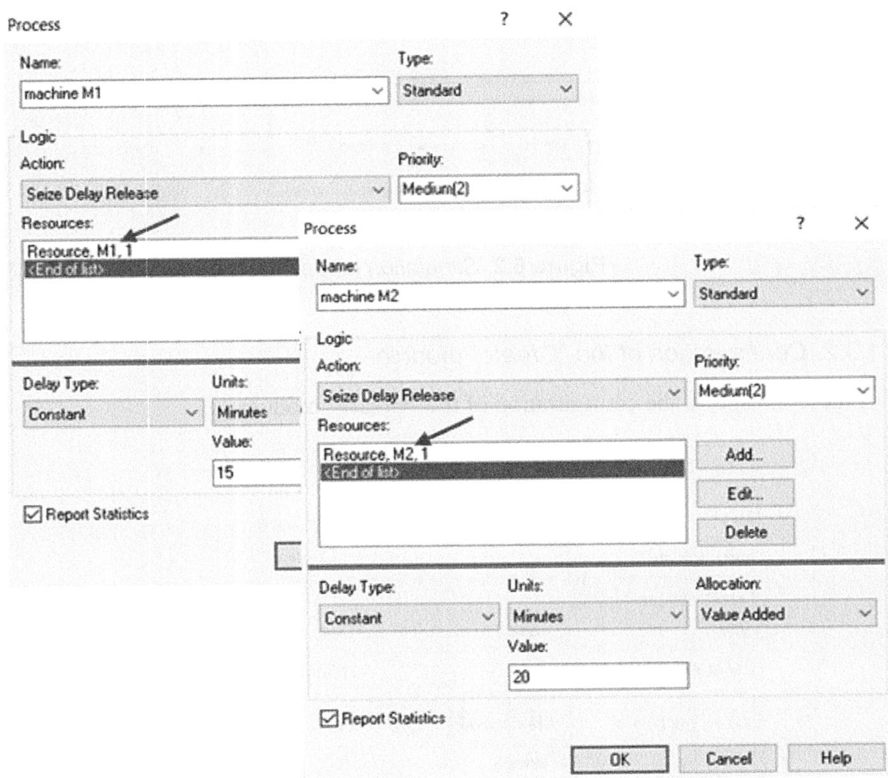

Figure 6.4. *Configuration of the "Process" modules "machine M1" and "machine M2"*

To visualize the queues of both machines, select the "Seize Delay Release" action type for the "Process" modules. Specify the processing times of 15 and 20 minutes for machines M_1 and M_2, respectively, in the "Delay Type" section of the "Process" modules, indicating that these times are constant. Make sure to check the "Report Statistics" box to generate the statistics for both machines in the final simulation report.

6.1.3.4. *Simulation stop condition*

A stop condition is required to halt arrivals at 100 parts. This condition is to be integrated through the "Run" > "Setup" directory. The stopping condition is defined as: "EntitiesIn(Entity 1) = 100" (Figure 6.5).

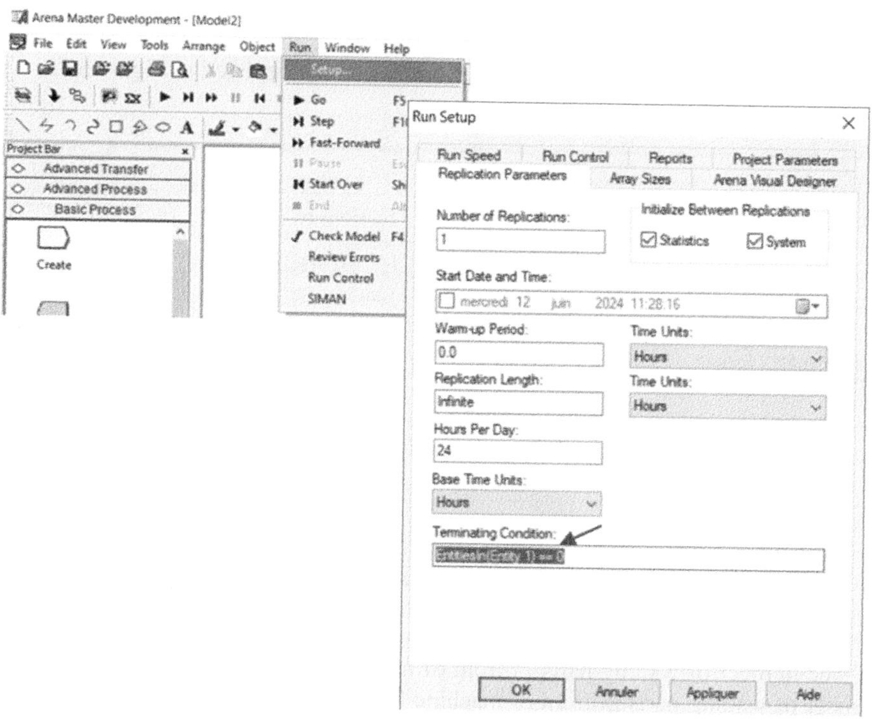

Figure 6.5. *Setting the simulation model stop condition*

6.1.4. *Comparison between results*

After running the simulation, the total time for the production of the 100 parts, as indicated at the bottom of the model, is 2,015 minutes (Figure 6.6). This aligns perfectly with the result obtained from manual simulation.

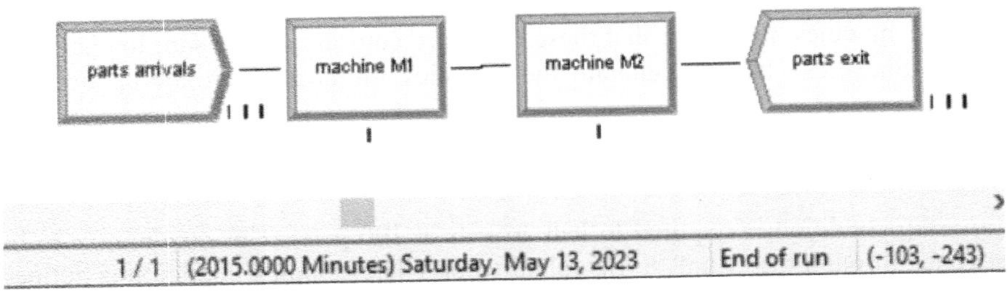

Figure 6.6. *Total simulation time*

The results obtained after running the simulation are depicted in Figure 6.7.

machine M1.Queue

Time	Average	Half Width	Minimum	Maximum
Waiting Time	0	(Insufficient)	0	0

Other	Average	Half Width	Minimum	Maximum
Number Waiting	0	(Insufficient)	0	0

machine M2.Queue

Time	Average	Half Width	Minimum	Maximum
Waiting Time	198.00	(Insufficient)	0	396.00

Other	Average	Half Width	Minimum	Maximum
Number Waiting	9.8263	(Insufficient)	0	20.0000

Figure 6.7. *Reporting simulation results*

The number of waiting parts in front of machine M_1 is zero, while the maximum number of waiting parts in front of machine M_2 is 20. This confirms the consistency between the Arena simulation and manual simulation results.

6.2. Collection and analysis of results (transitional period detection)

6.2.1. *Description and data*

The goal of this manipulation is to apply methods for collecting and analyzing the results of a stochastic simulation model. Every simulation model starts empty. As a result, the first observations cannot represent the normal operating conditions of the system. These observations, considered as a "Warm-up" period, must then be eliminated before collecting the steady state statistics because they represent a transient regime of the simulation model.

The considered manufacturing system is composed of three machines placed in series. The parts arrive at the system with an inter-arrival time following a Poisson distribution with a rate $\lambda = 6$ minutes. The processing times of the parts on each machine follow exponential distributions with averages of 4 minutes and 5 minutes, respectively, along with a uniform distribution ranging from 3 to 8 minutes.

We can model this manufacturing system using Arena software. There are two methods to analyze simulation results while avoiding unwanted effects from the warm-up period.

The first method consists of carrying out a long replication:

– perform a 100,000 minutes replication;

– create a graph illustrating the throughput of parts within the system. System throughput is determined as the ratio of produced parts to their exiting times;

– collect the average values and plot the evolution curves on their respective graphs for the average number of entities in the queues of the three machines, "NWM1", "NWM2" and "NWM3", along with the utilization rates of the machines, "UTILM1", "UTILM2" and "UTILM3".

The second method consists of carrying out "n" independent replications and eliminating the warm-up period, which can be detected from a control simulation:

– estimate the warm-up period from the previous graph of the throughput;

– carry out 10 replications of 10,000 minutes each while eliminating the warm-up period;

– collect the different performance indexes for each machine during each replica;

– determine, using Excel, the mean "X" and the standard deviation "S" of each performance index;

– calculate for each performance index the corresponding confidence interval $CI = \overline{X} \mp \frac{S}{\sqrt{N}} t_{\alpha/2,N-1}$ with $\alpha = 10\%$;

– compare the values of the performance indexes obtained from the first method to the CIs of the second method.

6.2.2. *Simulation model*

The simulation model is composed of a "Create" module called "product arrivals", followed by three "Process" modules placed in series, which are respectively named "machine M1", "machine M2" and "machine M3". An "Assign" module, called "throughput calculation", is subsequently used to compute the throughput rate of the manufacturing system. The simulation model must be concluded by a "Dispose" module called "parts exit" (Figure 6.8).

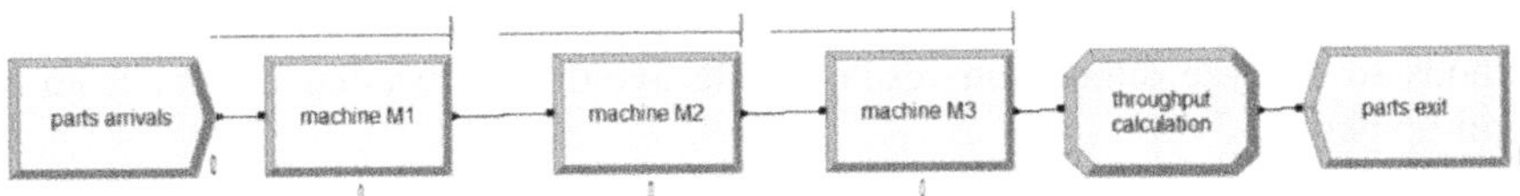

Figure 6.8. *Simulation model*

6.2.2.1. *Configuration of the "Create" module*

The inter-arrival time follows a Poisson distribution. Hence, it must be specified as the "Expression" type to access the different statistical distributions in the "Expression" box of the "Create" module. It is then sufficient to choose the type of distribution as "POIS" and to specify the value of its rate $\lambda = 6$ minutes (Figure 6.9).

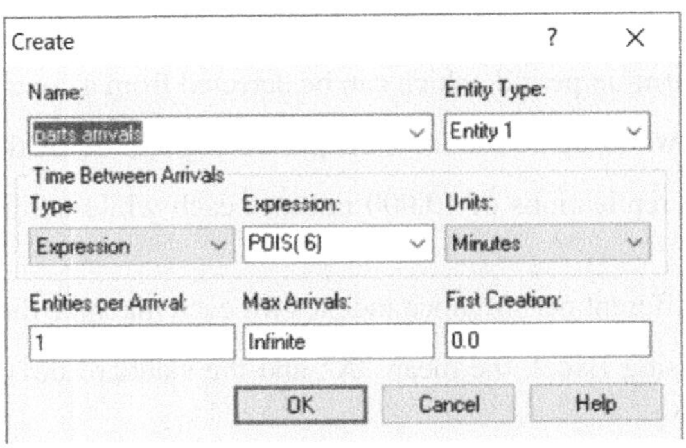

Figure 6.9. *Configuration of the "Create" module "product arrival"*

6.2.2.2. *Configuration of "Process" modules*

The part's processing times in the machines M_1, M_2 and M_3 follow statistical distributions, which must be indicated in the "Delay Type" section of the "Process" modules using the "Expression" type (Figure 6.10).

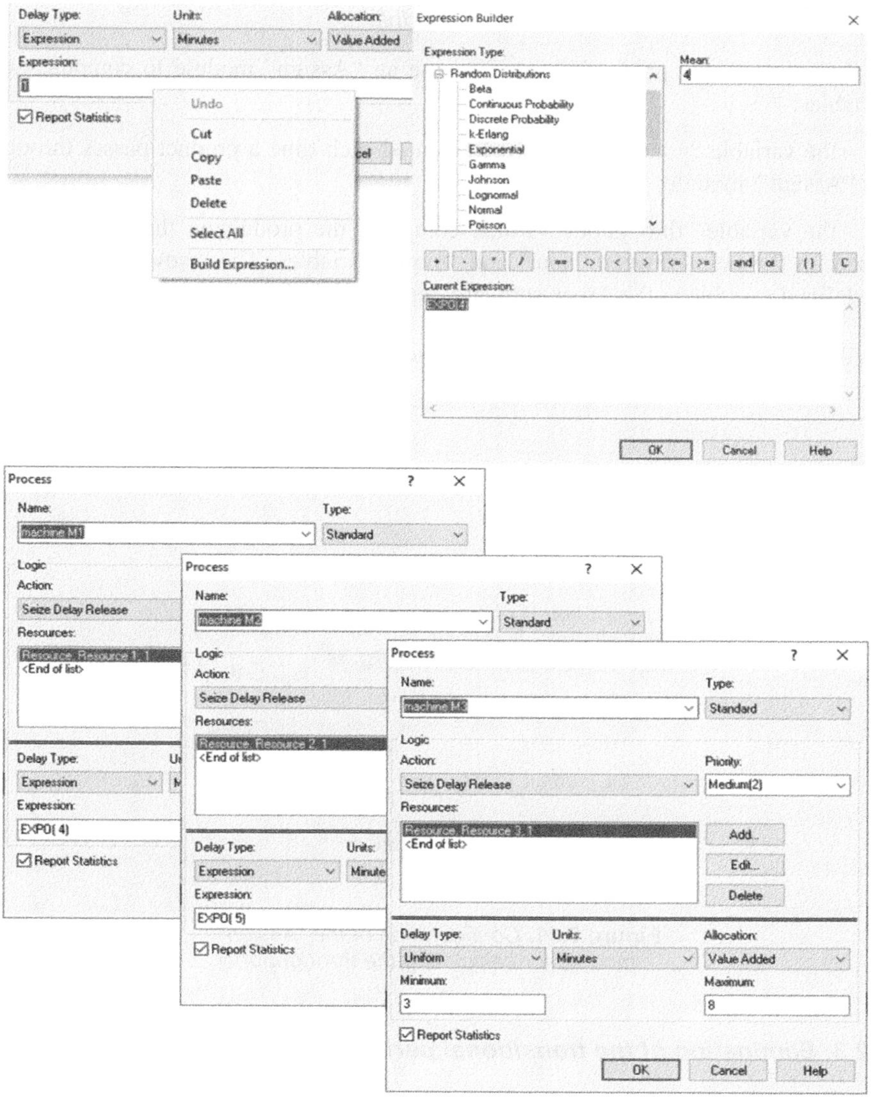

Figure 6.10. *Configuration of the "Process" modules "machine Mi"*

To specify the nature and characteristic values of the statistical distribution, we simply place the mouse pointer in the white area of the expression, press the right mouse button and choose the "Built" option in the new window that appears. The adequate distribution can be chosen using the "Random Distributions" option of the "Expression Builder" function.

6.2.2.3. *Configuration of the "Assign" module*

To calculate the production rate, we use an "Assign" module to compute two variables:

– the variable "nout", which is incremented each time a product passes through the "Assign " module;

– the variable "throughput", which computes the production throughput value based on the ratio between "nout" and "tnow" variables. The "tnow" variable is a predefined variable in the Arena software, representing the current simulation time.

We use the "Add" box in the "Assign" module to add a variable (Figure 6.11).

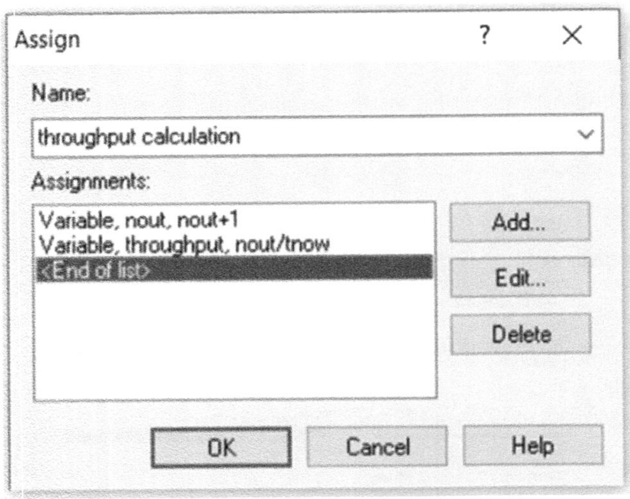

Figure 6.11. *Configuration of the "Assign" module for calculating the throughput*

6.2.3. *Elimination of the transitional period*

6.2.3.1. *Method 1: large replication*

Follow the "Run" > "Setup" directory to program a single replica of 100,000 minutes (Figure 6.12).

To create the throughput graph, we use the "PLOT" icon to specify the expression of the variable to plot, add series of data in the "Data Series" section by pressing the "Add" button and select the "throughput" variable using the elevator in the "Expression" box of the "Source Data" section (Figure 6.13).

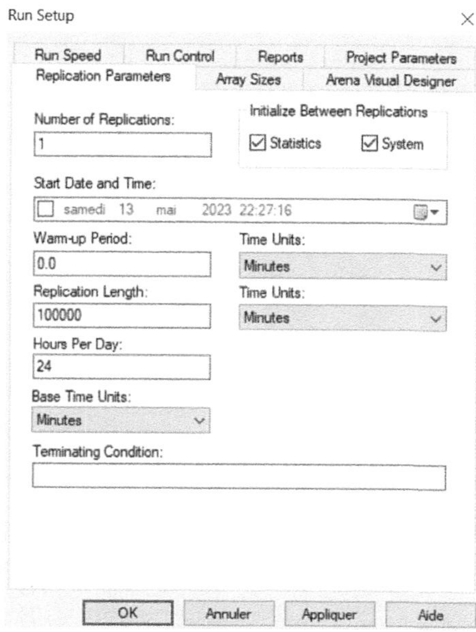

Figure 6.12. *Setting the simulation length*

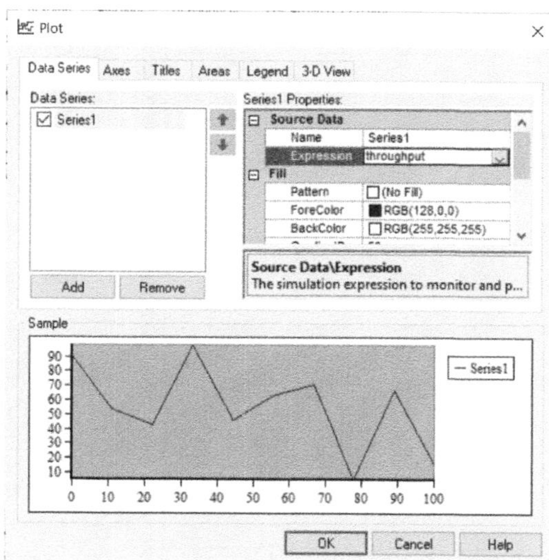

Figure 6.13. *Setting the "PLOT" module to plot the throughput*

The obtained graph for the system throughput is as follows (Figure 6.14).

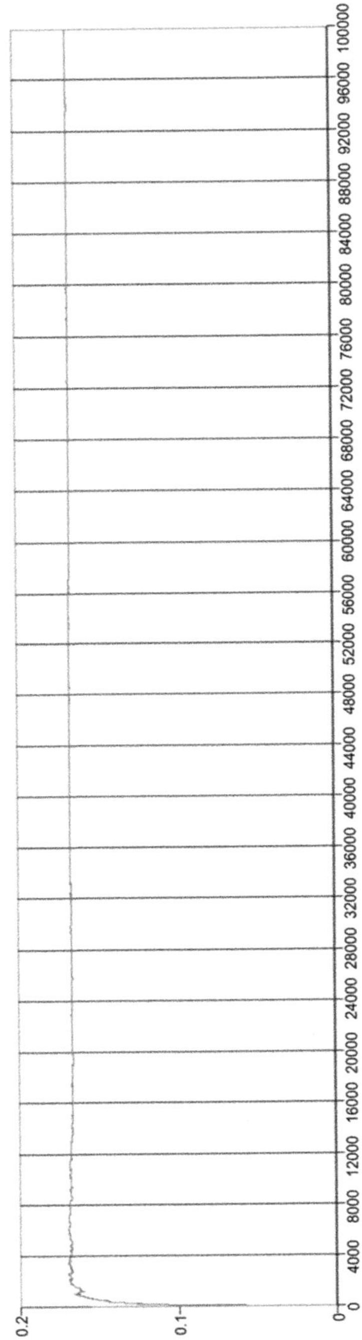

Figure 6.14. *Graph of the throughput of the system*

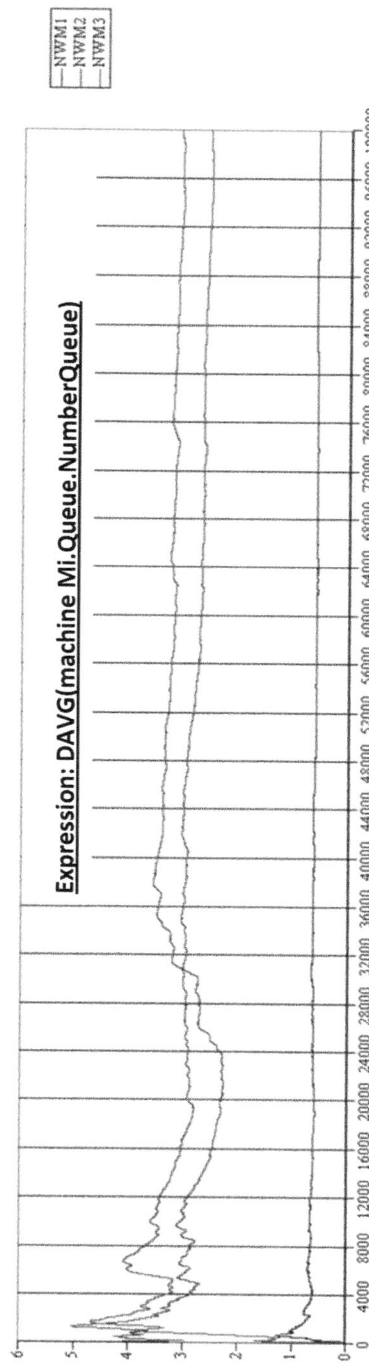

Figure 6.15. *Graph of the average number of entities in the queues of the three machines. For a color version of the figure, see www.iste.co.uk/jerbi/discrete.zip*

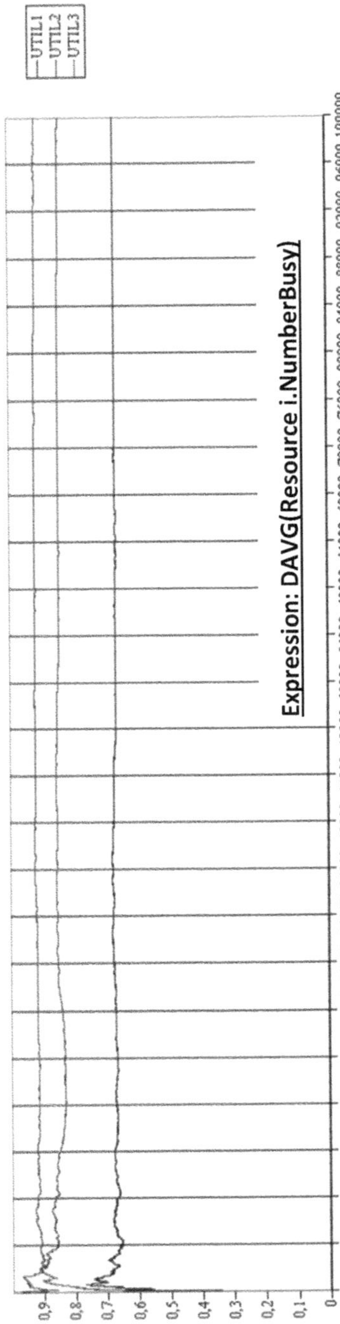

Figure 6.16. *Graph of the utilization rate of the three resources. For a color version of the figure, see www.iste.co.uk/jerbi/discrete.zip*

In the case of the average numbers of entities in the queues of the three machines "NWM1", "NWM2" and "NWM3" and the utilization rates of the machines "UTILM1", "UTILM2" and "UTILM3", we use the "Build Expression" tool in the "Expression" part of the "Source Data" section of the "PLOT" module (Figures 6.15 and 6.16).

The simulation results are as follows (Figures 6.17 and 6.18).

Usage

	Inst Util	Num Busy	Num Sched	Num Seized	Sched Util
Resource 1	0.67	0.67	1.00	16 619.00	0.67
Resource 2	0.84	0.84	1.00	16 618.00	0.84
Resource 3	0.91	0.91	1.00	16 615.00	0.91

Figure 6.17. *Resource usage*

machine M1.Queue

Time	Average	Half Width	Minimum	Maximum
Waiting Time	3.6601	0.415028327	0	54.8337

Other	Average	Half Width	Minimum	Maximum
Number Waiting	0.6083	0.061344613	0	10.0000

machine M2.Queue

Time	Average	Half Width	Minimum	Maximum
Waiting Time	15.4664	(Correlated)	0	131.93

Other	Average	Half Width	Minimum	Maximum
Number Waiting	2.5702	(Correlated)	0	25.0000

machine M3.Queue

Time	Average	Half Width	Minimum	Maximum
Waiting Time	18.7310	2.38935	0	107.77

Other	Average	Half Width	Minimum	Maximum
Number Waiting	3.1125	0.404581769	0	20.0000

Figure 6.18. *Number of waiting entities*

6.2.3.2. *Method 2: n replications*

We use the “Run” > “Setup” directory to program the 10 replications of 10,000 minutes.

Estimation of the warm-up period: according to the resource usage graph, the system reaches its permanent state after 2,000 minutes (Figure 6.19).

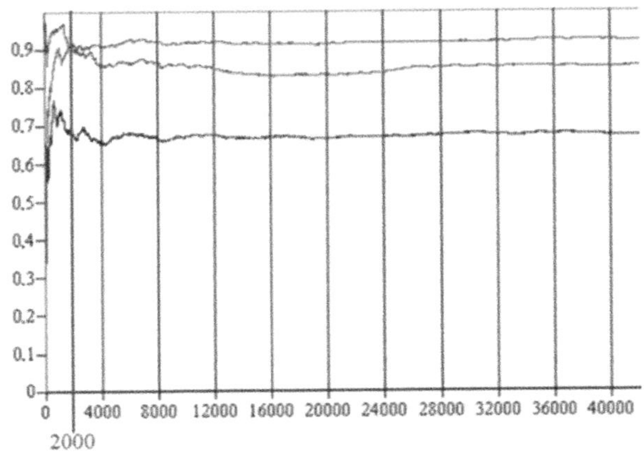

Figure 6.19. *Determination of the transitional period. For a color version of the figure, see www.iste.co.uk/jerbi/discrete.zip*

Table 6.1 presents the result of the calculated mean “X” and standard deviation “S” of each performance index, along with their corresponding CI.

Replication	NWM1	NWM2	NWM3	UTILM1	UTILM2	UTILM3
1	0.655	2.886	3.888	0.670	0.845	0.919
2	0.517	3.148	2.886	0.657	0.800	0.914
3	0.550	1.834	2.829	0.679	0.871	0.933
4	0.626	2.936	3.611	0.653	0.806	0.899
5	0.650	2.305	2.428	0.680	0.805	0.917
6	0.620	1.932	2.793	0.681	0.838	0.917
7	0.614	2.384	2.575	0.676	0.850	0.921
8	0.670	2.600	3.373	0.682	0.831	0.915

Replication	NWM1	NWM2	NWM3	UTILM1	UTILM2	UTILM3
9	0.736	2.576	2.542	0.644	0.834	0.906
10	0.607	2.825	2.451	0.656	0.845	0.923
Average	0.625	2.543	2.938	0.668	0.833	0.916
Standard deviation	0.061	0.432	0.513	0.014	0.023	0.009
IC	0.032	0.225	0.267	0.007	0.012	0.005
IC lower terminal	0.593	2.318	2.671	0.661	0.821	0.912
IC upper terminal	0.656	2.767	3.204	0.675	0.844	0.921

Table 6.1. *Results of simulation replicates*

Comparing the two simulation methods reveals that the performance index value obtained using the large replication method falls within the CIs obtained by the n replication method. This highlights the significance of selecting the warm-up period in the latter method to mitigate its effects without affecting the precision of the results (Figure 6.20).

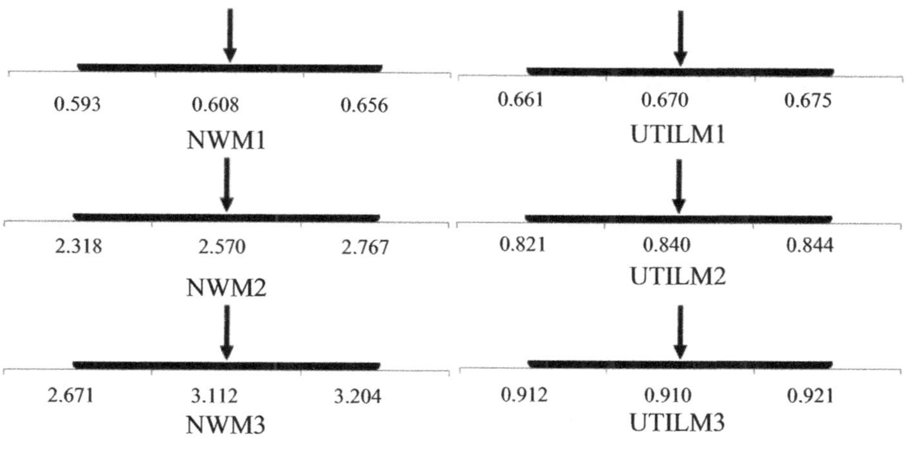

Figure 6.20. *Comparison between the results of the two methods of eliminating the transitional period*

6.3. Resource failure

6.3.1. *Description and data*

The aim of this manipulation is to master the parameterization of a stochastic simulation model by introducing the notion of the breakdown of the resources used.

Citizens enter a postal center with a single counter. They arrive individually according to an inter-arrival time that follows an exponential distribution with an average of 3 minutes. The processing time for these citizens' requests follows a triangular distribution of T(1, 3, 6) minutes.

(1) We develop a simulation model of a postal center. We run the simulation model with a replication of length 480 minutes and collect the following performance indexes:

– the number N of citizens whose requests have been processed;

– the counter utilization rate;

– the average waiting time at the counter;

– the average number of waiting citizens;

– the mean flow time (MFT);

– the work in process (WIP).

(2) We run two replications of 480 minutes each. What do you observe from the simulation results of these two replications? Can you explain why?

(3) For a single replication of 480 minutes, we plot the graph of the number of citizens waiting in front of the counter as well as the graph presenting the evolution of the states of use of the counter. Remember that if the counter is occupied, its state = 1, and if it is free, its state = 0.

(4) In reality, the post office's computer network breaks down. The time between two consecutive breakdowns, "Up time", follows an exponential distribution with an average of 72 minutes and the network restoration time, "Down time", follows a uniform distribution between 4 and 15 minutes. During the network restoration time, the current citizen must wait to:

– introduce the breakdowns into the model, run a 480 minute simulation and collect the performance indexes;

– consult the simulation report entitled "Frequencies" and give the total duration of breakdowns and the percentage of times when the counter is free.

6.3.2. *Simulation model*

The simulation model is quite simple. It consists of a "Create" module, a "Process" module and a "Dispose" module (Figure 6.21).

Figure 6.21. *Simulation model*

After simulating a single 480 minute replication, the following results are obtained:

– the number N of citizens whose requests were processed = 140 citizens;

– the counter utilization rate = 0.986%;

– the average waiting time in front of the counter = 18.023 minutes;

– the average number of citizens waiting = 5.350 citizens;

– the MFT = 21.439 minutes;

– the WIP = 6.339 citizens.

6.3.3. *Replications*

The simulation results are presented in Figures 6.22 and 6.23 for two replications. Notably, there are differences between the results of these two replications, attributable to the stochastic nature of the simulation model under examination.

Consequently, the results obtained for different replications can then be considered as measurements on a statistical sample. In order to effectively use these simulation results in a decision-making process, it is essential to conduct appropriate and thorough statistical analysis.

Replication 1 Start Time: 0.00 Stop Time: 480.00 Time Units: Minutes

Entity

Time

VA Time	Average	Half Width	Minimum	Maximum
Entity 1	3.3800	(Insufficient)	1.3536	5.8582

NVA Time	Average	Half Width	Minimum	Maximum
Entity 1	0	(Insufficient)	0	0

Wait Time	Average	Half Width	Minimum	Maximum
Entity 1	18.0594	(Insufficient)	0	45.7321

Other

Number In	Value
Entity 1	144

Number Out	Value
Entity 1	140

WIP	Average	Half Width	Minimum	Maximum
Entity 1	6.3394	(Insufficient)	0	14.0000

Queue

Time

Waiting Time	Average	Half Width	Minimum	Maximum
post office counter.Queue	18.0234	(Insufficient)	0	45.7321

Resource

Usage

Instantaneous Utilization	Average	Half Width	Minimum	Maximum
counter	0.9864	(Insufficient)	0	1.0000

System

Other

Number Out	Value
System	140

Figure 6.22. *First replication simulation results*

Replication 2 Start Time: 0.00 Stop Time: 480.00 Time Units: Minutes

Entity

Time

NVA Time	Average	Half Width	Minimum	Maximum
Entity 1	0	(Insufficient)	0	C

Wait Time	Average	Half Width	Minimum	Maximum
Entity 1	13.3673	(Insufficient)	0	32.0207

Other

Number In	Value
Entity 1	150

Number Out	Value
Entity 1	139

WIP	Average	Half Width	Minimum	Maximum
Entity 1	5.2061	(Insufficient)	0	14.000C

Queue

Time

Waiting Time	Average	Half Width	Minimum	Maximum
post office counter.Queue	13.4720	(Insufficient)	0	32.0207

Resource

Usage

Instantaneous Utilization	Average	Half Width	Minimum	Maximum
counter	0.9611	(Insufficient)	0	1.000C

System

Other

Number Out	Value
System	139

Figure 6.23. *Second replication simulation results*

6.3.4. *Graphs*

To plot the graphs, we use the "PLOT" icon, specifying the expression to be plotted. It is then a question of plotting "NQ(post office counter.Queue)" (Figure 6.24) for the graph of the number of citizens waiting and plotting "NR(box office)"

(Figure 6.25) for the graph presenting the evolution of the states of use of the counter resource.

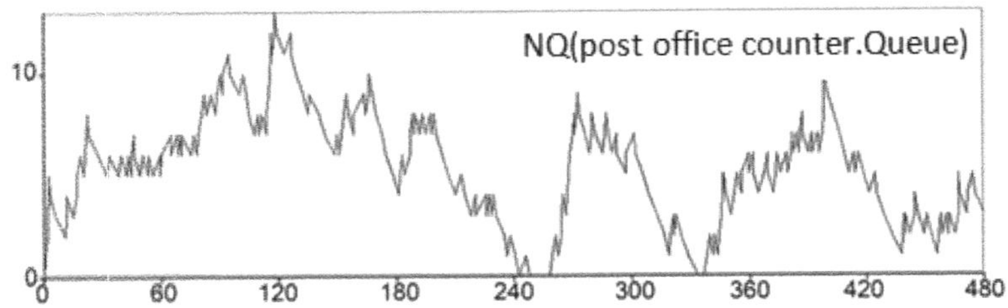

Figure 6.24. *Graph of the number of entities in the queue*

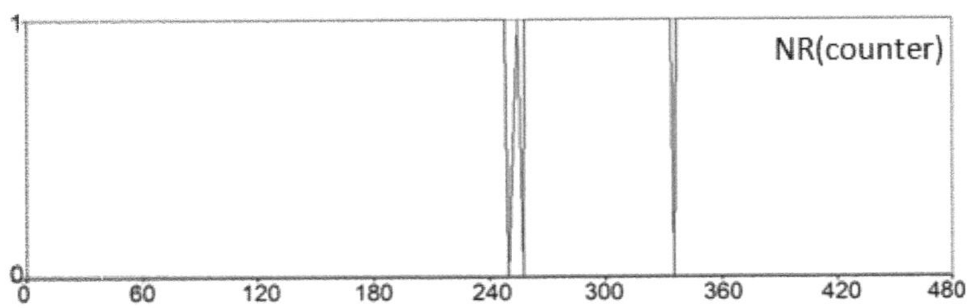

Figure 6.25. *Counter resource status graph*

6.3.5. *Breakdowns*

The breakdowns are defined as failures in the "Failure" module located in the "Advanced process" library. It is then a matter of specifying the parameters of "Failure 1", such as the times between the "Up Time" failures and the "Down Time" failure times (Figure 6.26).

Failure - Advanced Process

	Name	Type	Up Time	Up Time Units	Down Time	Down Time Units	Uptime in this State only
1	Failure 1	Time	72	Minutes	UNIF(4, 15)	Minutes	

Figure 6.26. *Setting the breakdowns*

This "Failure 1" is subsequently assigned to the "counter" resource with a "Wait" -type stop in the "Resource" module of the "Basic Process" library to respect the condition "During the network restoration time, the current citizen must wait for" (Figure 6.27).

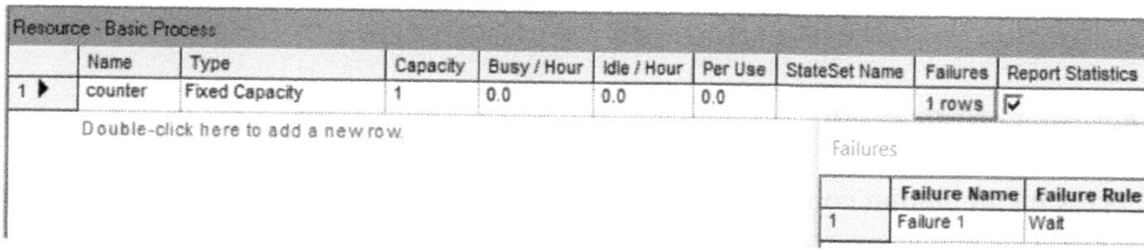

Resource - Basic Process

	Name	Type	Capacity	Busy / Hour	Idle / Hour	Per Use	StateSet Name	Failures	Report Statistics
1	counter	Fixed Capacity	1	0.0	0.0	0.0		1 rows	☑

Double-click here to add a new row.

Failures

	Failure Name	Failure Rule
1	Failure 1	Wait

Figure 6.27. *Setting up the resource*

The collection of the "Frequency" of the "state" of the resource ("Busy", "Idle", "Failed") must be configured in the "Statistic" module located in the "Advanced process" library (Figure 6.28)

Statistic - Advanced Process

	Name	Type	Frequency Type	Resource Name	Collection Period	Report Label	Output File	Categories
1	Statistic 1	Frequency	State	counter	Entire Replication	Statistic 1		0 rows

Figure 6.28. *Configuration of the "Statistic" module for the resource*

After simulating a single 480 minute replication, the following results are obtained:

– the number N of citizens whose requests were processed = 133 citizens;

– the window utilization rate = 0.925;

– the average waiting time in front of the counter = 26.998 minutes;

– the average number of citizens waiting = 8.156 citizens;

– the MFT = 30.336 minutes;

– the WIP = 9.081 citizens.

During the 480 minutes, the counter went down six times with an average waiting time of 6.008 minutes, which makes a total of 7.51% of the simulation time spent waiting for the network to recover. The counter was then used for the remaining 92.49% of the time (Figure 6.29).

Statistic 1	Number Obs	Average Time	Standard Percent	Restricted Percent
BUSY	6	73.9912	92.49	92.49
FAILED	6	6.0088	7.51	7.51

Figure 6.29. *Resource statistics results*

6.4. Entities transfer (PickStation)

6.4.1. *Description and data*

This manipulation consists of developing a simulation model which evolves over several stages with the introduction of the use of the " PickStation" module.

(1) A mobile phone repair company consists of a formatting and updating workstation and two parallel repair workstations. A reception desk receives telephones from customers who drop them off every 20 minutes. The reception process takes 3 minutes, while the repair process takes 25 minutes, and the formatting and updating process takes 15 minutes. Once received, each phone is sent to the repair workstation with the lowest number of phones waiting, then to the formatting and updating workstation if necessary. Given that all phones arriving at the repair company require repair, formatting and updating, we simulate the operation of this company to deal with 20 phones. We determine the required simulation time.

(2) Field observations indicate that customers leave their phones according to an inter-arrival time following an exponential distribution with an average of 14 minutes. In addition, the repair time follows an exponential distribution with an average of 25 minutes while the minimum and maximum formatting and updating times are respectively 12 and 15 minutes. Furthermore, it is noted that only 80% of repaired phones require formatting and updating, while the remaining 20% require only repair. We update the simulation model, simulate the operation of this company for 8 hours and measure the following indexes:

– the MFT of the telephones in the company;

– the average number of telephones waiting in front of each workstation;

– the activity rate of the different workstations.

(3) Plot the graphs representing the number of phones waiting in front of the repair workstations and in front of the formatting and updating workstation on a single graph. To enhance its service, the company introduces a tablet repair, formatting and updating services. The updated system data are then summarized in Table 6.2.

Type	Inter-arrival times (minutes)	Repair time (minutes)	Format and update time (minutes)
Phone	EXPO (14)	EXPO (25)	UNIF [12.15]
Tablet	EXPO (60)	EXPO (45)	UNIF [20.36]

Table 6.2. *System data*

Given that the percentage of phones that require repairs and formatting remains unchanged, while only half of the repaired tablets require formatting and updating afterward:

– modify the simulation model and simulate the operation of this business for 8 hours;

– collect the performance indexes of this system;

– what observations can be made? Suggest a solution.

6.4.2. *Simulation model*

The selection of the repair workstations is determined by the number of phones waiting at each workstation. Therefore, the decision is based on choosing the workstation with the lowest number of telephones in queue. This selection can be modeled in two ways:

– by using a "Decide" module of the "2-way by Condition" type, which will be integrated into the output of the reception counter and which is based on the equation "NQ(repair1.Queue) < NQ(repair2.Queue)", to check whether the number of phones waiting in the queue of "repair1" is less than that waiting in the queue of "repair2" (Figures 6.30 and 6.31);

– by using the "PickStation" module coupled with two "Station" modules each connected to one of the repair workstations. This module makes it possible to choose the workstation with the lowest number of phones in queue. To achieve this, the "Number in Queue" box must be checked by specifying the "Minimum" value when configuring the "PickStation" module test. It is then necessary to specify the stations from which the selection will be made. Due to the selection condition, the stations are connected to queues in the "Stations" section of the "PickStation" module. Therefore, specify "Station1" and "Station 2" along with the respective resource queues connected to them, "repair1.Queue" and "repair2.Queue" (Figures 6.32 and 6.33).

In both cases, it appears that the simulation time required to process the 20 phones is 443 minutes.

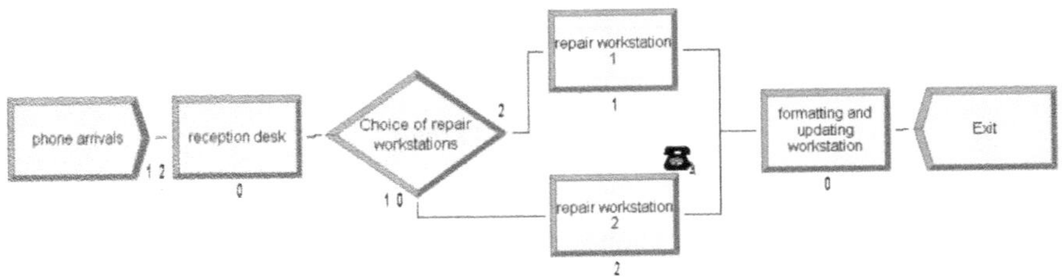

Figure 6.30. *First version simulation model*

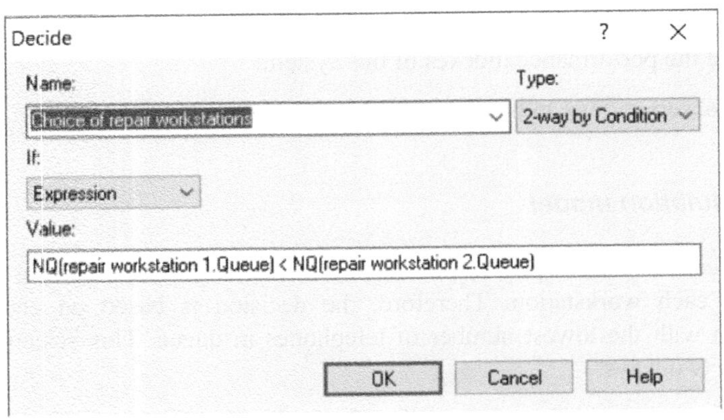

Figure 6.31. *Configuration of the "Decide" module*

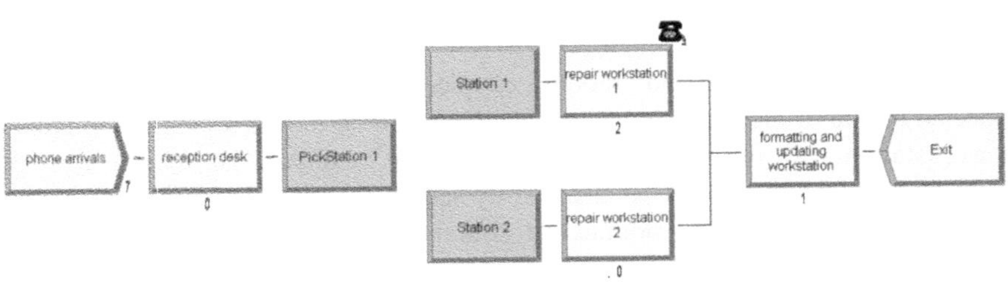

Figure 6.32. *Second version simulation model. For a color version of the figure, see www.iste.co.uk/jerbi/discrete.zip*

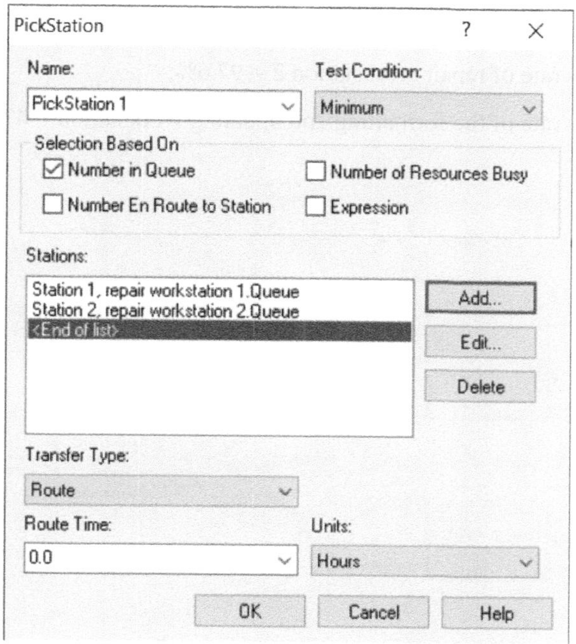

Figure 6.33. *Configuration of the "PickStation" module*

6.4.3. *Choice of operation (formatting)*

A new "Decide" module, called "formatting?", must be integrated into the simulation model. This module will be placed just after the two repair workstations. It is of the "2-way by chance" type and sorts the phones based on a percentage condition. Indeed, 80% of the phones will be directed to the "formatting and updating" workstation, while the remaining 20% will be directed to the simulation model exit (Figures 6.34 and 6.35).

The simulation results of an 8 hour working day are as follows:

– the MFT = 114.40 minutes;

– the average number of phones waiting in front of repair workstation 1 = 3.947 phones;

– the average number of phones waiting in front of repair workstation 2 = 3.861 phones;

– the average number of phones waiting in front of the formatting and updating workstation = 0.123 phones;

– the activity rate of repair workstation 1 = 99.4%;

– the activity rate of repair workstation 2 = 97.6%;

– the activity rate of the formatting and updating workstation = 47.5%.

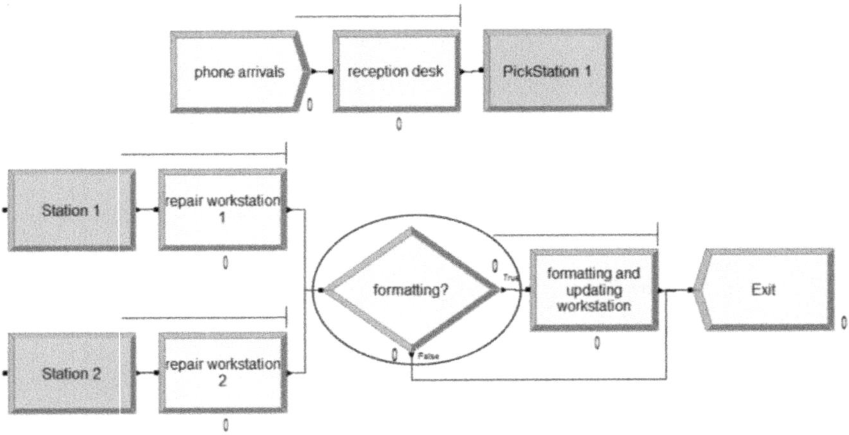

Figure 6.34. *Upgrade simulation model. For a color version of the figure, see www.iste.co.uk/jerbi/discrete.zip*

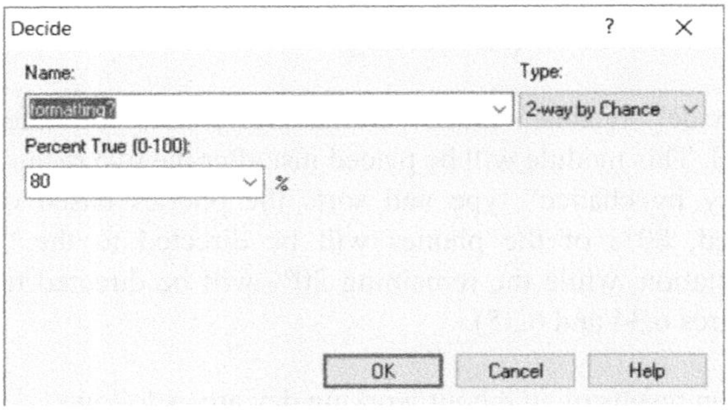

Figure 6.35. *Configuration of the "Decide" module "formatting?"*

6.4.4. *New product (tablet)*

To model the arrivals of the tablets, a "Create" module must be added to the simulation model. In addition, for each type of entity, tablet and phone, an "Assign" module must be added to assign its processing times in each workstation (Figure 6.36).

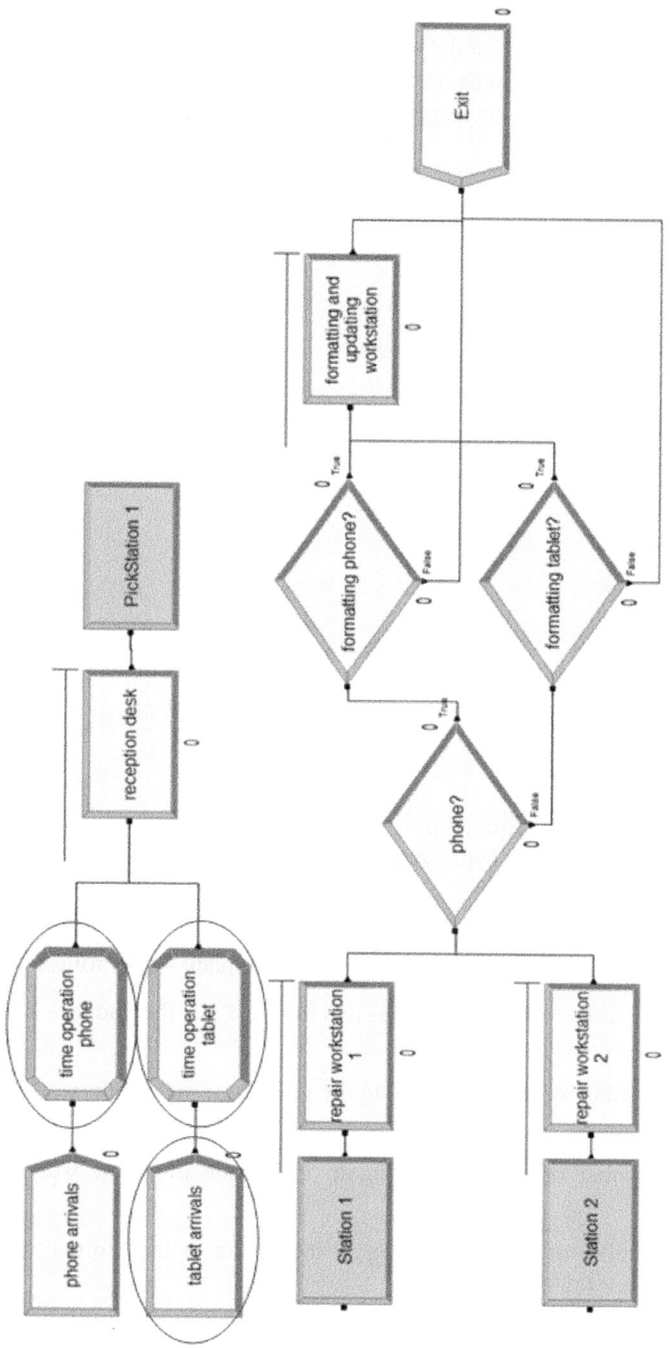

Figure 6.36. *Simulation model with tablet. For a color version of the figure, see www.iste.co.uk/jerbi/discrete.zip*

The configuration of repair, formatting and updating times can no longer be configured within the "Process" modules, as these times vary depending on the entity type. Thus, these times must be assigned as the attributes "TREP" and "TFOR" to each type of entity in the respective "Assign" modules (Figure 6.37).

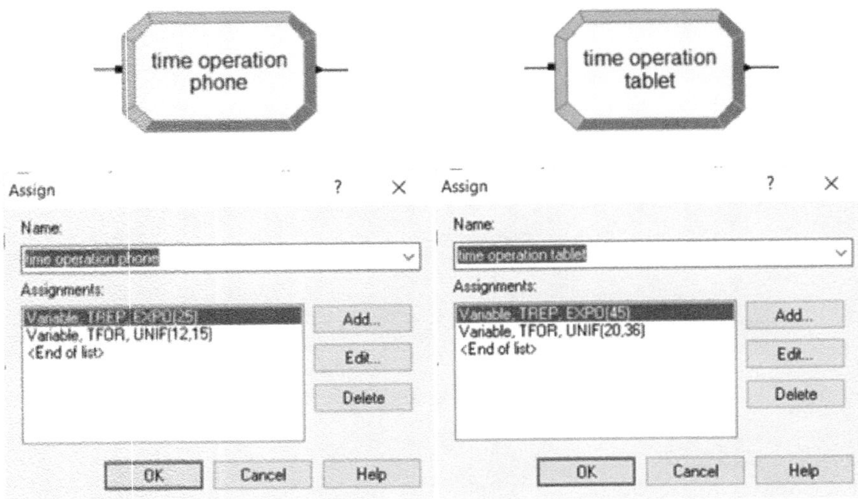

Figure 6.37. *Configuration of "Assign" modules "time operation phone" and "time operation table"*

These attributes are then integrated into each corresponding "Process" module in the expression part of the "Delay Type" zone (Figure 6.38).

The simulation results are as follows:

– the number of phones and tablets handled by the company = 25 entities;

– the MFT for the tablet = 83.472 minutes and the MFT for the phone = 128.280 minutes;

– the average number of entities waiting in the queue of repair workstation 1 = 3.982 entities;

– the average number of entities waiting in the queue of repair workstation 2 = 3.667 entities;

– the average number of entities waiting in the queue of the "formatting and updating" workstation = 0.299 entities;

– the activity rate of the "repair1" workstation= 91.5%;

– the activity rate of the “repair2” workstation = 98.1%;

– the activity rate of the “formatting and updating” workstation = 59.2%.

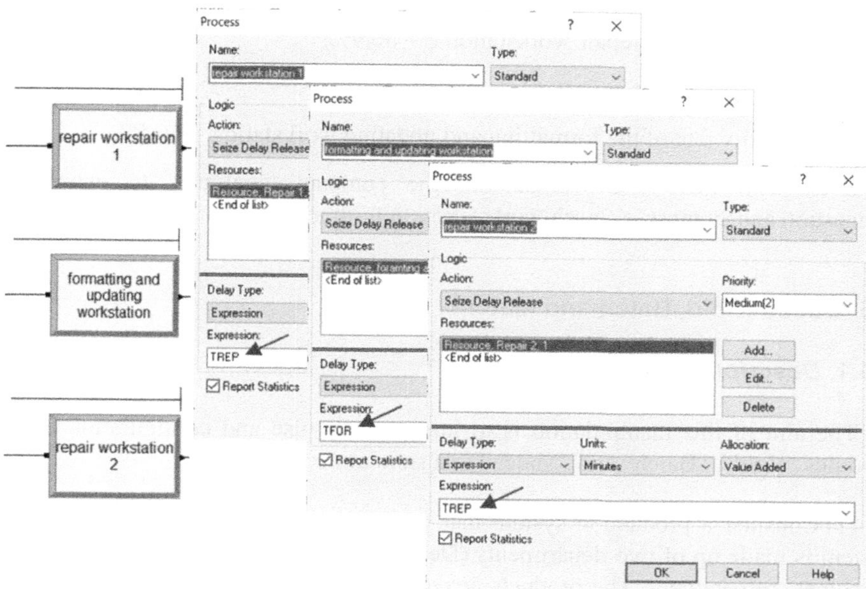

Figure 6.38. *Configuration of “Process” modules with attribute*

The system is experiencing congestion, evident from the high utilization rate of the workstations and the average number of entities waiting. Adding a third repair workstation would be a strategic move, particularly considering the significant demand for repair services. The updated simulation results with three repair workstations are as follows:

– the number of telephones and tablets processed by the company = 35 entities;

– the MFT of the tablet = 63.039 minutes and the MFT of the phone = 87.380 minutes;

– the average number of entities waiting in front of repair workstation 1 = 1.154 entities;

– the average number of entities waiting in front of repair workstation 2 = 1.056 entities;

– the average number of entities waiting in front of repair workstation 3 = 0.666 entities;

– the average number of entities waiting in front of the "formatting and updating" workstation = 0.3945 entities;

– the activity rate of repair workstation 1 = 90.2%;

– the activity rate of repair workstation 2 = 80.9%;

– the activity rate of repair workstation 3 = 76.7%;

– the activity rate of the formatting and updating workstation = 66.5%.

Even with this third workstation, the company continues to experience congestion and supports a fourth workstation to further decongest its system.

6.5. Queue (Hold, Batch and Match)

6.5.1. *Description and data*

The aim of this manipulation is to introduce the use and configuration of the modules "Hold", "Batch" and "Match" in a simulation model.

(1) Consider a production system that processes a single type of product. This system is made up of two departments (Dep_1 and Dep_2) composed of two and three machines, respectively. The products arrive at the production system in batches of three units according to an inter-arrival time which is unknown, but the only known data are a minimum (3 minutes) and a maximum (5 minutes).

(2) Once created, the products are transferred to Dep_1 then to Dep_2. When they arrive at one of the departments, the products are put on hold in a queue belonging to the department waiting for one of its machines to become free. In this case, the first waiting product is transferred to this machine.

(3) The production times necessary for the processing of products in Dep_1 and Dep_2 are all constant and are respectively equal to 4 and 12 minutes. It is assumed that the times necessary for the transfer of products between the different departments of the system are all zero:

– model and simulate the operation of this system for 1,000 minutes;

– estimate the warm-up period and eliminate it from the statistics, then collect the results for two replications.

(4) At the end of this production system, quality control and packaging workstations are added. The control time follows a uniform distribution between 3 and 4 minutes. After the control, the products are packaged in packs of two units. The packing time of a package follows the normal distribution with an average of 1 minute

and a standard deviation of 0.5 minutes. It is assumed that the empty boxes required for packaging are always available at the packing workstation:

– develop the model and simulate the operation of this workshop for a period of 1,000 minutes;

– collect the simulation result.

(5) In reality, the empty boxes required for packaging arrive at the packing workstation according to an exponential distribution with an average of 8 minutes. We update the simulation model.

6.5.2. *Simulation model (Hold)*

Figure 6.39 illustrates the simulation model to develop.

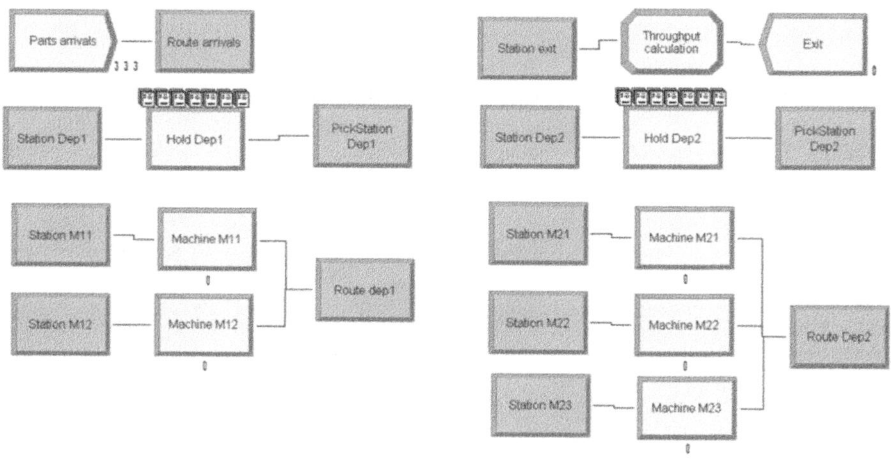

Figure 6.39. *Simulation model. For a color version of the figure, see www.iste.co.uk/jerbi/discrete.zip*

Considering the unknown inter-arrival times of the products and only the minimum of 3 minutes and maximum of 5 minutes being known, the inter-arrival time is set to follow a uniform distribution UNIF [3, 5] minutes when configuring the “Create” module named “product entry”. It is important to specify that the “Entities per arrival” = 3 as the products arrive in batches of 3 (Figure 6.40).

The queues of the two departments are modeled by two "Hold" modules. These modules retain the products until the conditions assigned to each are fulfilled. These conditions are designed to monitor the WIP status of the machines within each department. When the WIP for a machine reaches zero, indicating it is available for further processing, the first product in the "Hold" queue is released (Figure 6.41).

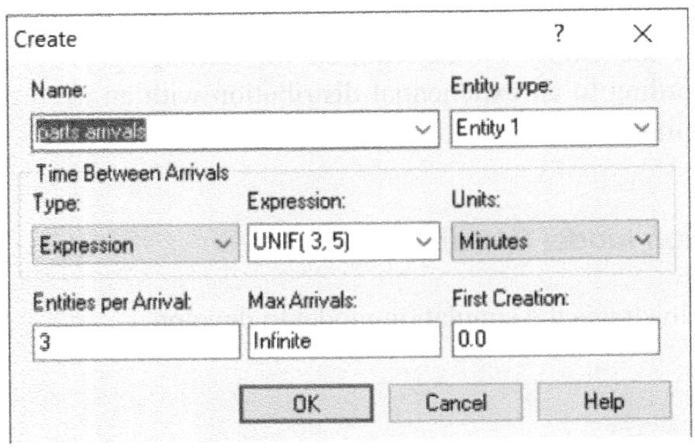

Figure 6.40. *Configuration of the "Create" module "product entry"*

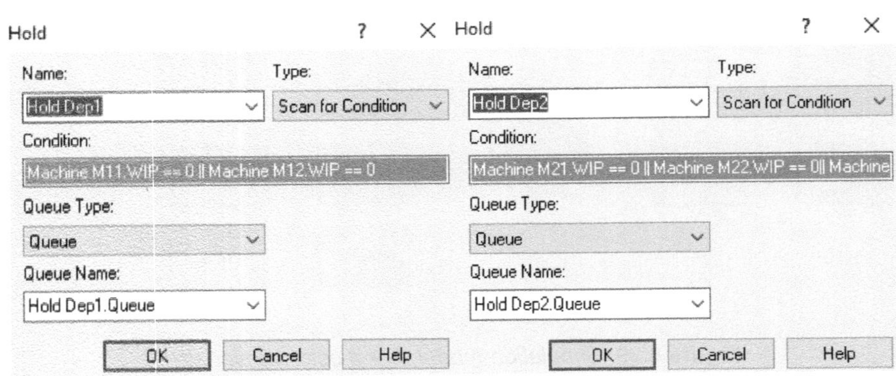

Figure 6.41. *Configuration of the "Hold", "Hold Dep1" and "Hold Dep2" modules*

The throughput rate graph shows that this rate stabilizes from 100 minutes, which then corresponds to the warm-up period (Figure 6.42).

The simulation model is run again for two replications while considering this warm-up period and the results are reproduced in Figures 6.43 and 6.44.

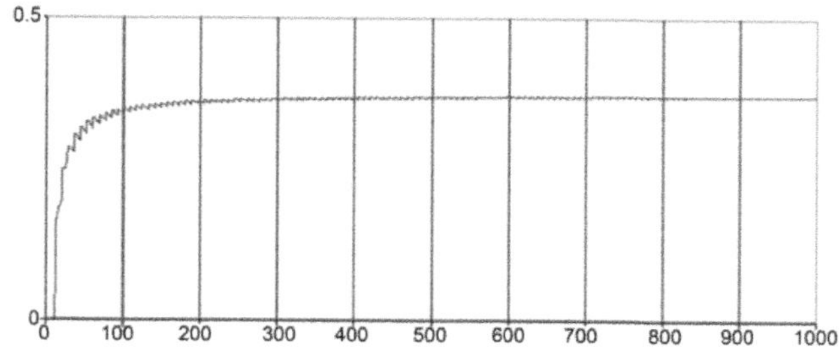

Figure 6.42. *Throughput graph*

Number In	Value			
Entity 1	675			
Number Out	**Value**			
Entity 1	227			
WIP	**Average**	**Half Width**	**Minimum**	**Maximum**
Entity 1	275.30	(Correlated)	52.0000	504.0C
Waiting Time	**Average**	**Half Width**	**Minimum**	**Maximum**
Hold Dep1.Queue	178.52	(Correlated)	33.5948	327.04
Hold Dep2.Queue	270.33	(Insufficient)	48.0000	492.0C
Number Waiting	**Average**	**Half Width**	**Minimum**	**Maximum**
Hold Dep1.Queue	133.96	(Correlated)	23.0000	250.0C
Hold Dep2.Queue	136.33	(Insufficient)	24.0000	251.0C
Instantaneous Utilization	**Average**	**Half Width**	**Minimum**	**Maximum**
M11	1.0000	(Insufficient)	0	1.000C
M12	1.0000	(Insufficient)	0	1.000C
M21	1.0000	(Insufficient)	0	1.000C
M22	1.0000	(Insufficient)	0	1.000C
M23	1.0000	(Insufficient)	0	1.000C

Figure 6.43. *Simulation results for replication 1*

Number In	Value			
Entity 1	672			
Number Out	**Value**			
Entity 1	227			
WIP	**Average**	**Half Width**	**Minimum**	**Maximum**
Entity 1	277.95	(Correlated)	52.0000	501.00
Waiting Time	**Average**	**Half Width**	**Minimum**	**Maximum**
Hold Dep1.Queue	181.38	(Correlated)	29.3044	332.30
Hold Dep2.Queue	270.33	(Insufficient)	48.0000	492.00
Number Waiting	**Average**	**Half Width**	**Minimum**	**Maximum**
Hold Dep1.Queue	136.62	(Correlated)	23.0000	248.00
Hold Dep2.Queue	136.33	(Insufficient)	24.0000	251.00
Instantaneous Utilization	**Average**	**Half Width**	**Minimum**	**Maximum**
M11	1.0000	(Insufficient)	0	1.0000
M12	1.0000	(Insufficient)	0	1.0000
M21	1.0000	(Insufficient)	0	1.0000
M22	1.0000	(Insufficient)	0	1.0000
M23	1.0000	(Insufficient)	0	1.0000

Figure 6.44. *Simulation results for replication 2*

6.5.3. *Quality control and packaging (Batch)*

The quality control step, incorporated at the end of the model, is represented by a "Process" module.

In addition, before proceeding to the packaging, parts need to be grouped into batches of two units. This is achieved using a "Batch" module, which holds each part until another arrives, grouping them into a batch to be sent to the "Process" module "packaging".

The packaging process is characterized by a duration following the normal distribution with an average of 1 minute and a standard deviation of 0.5 minutes (Figures 6.45 and 6.46).

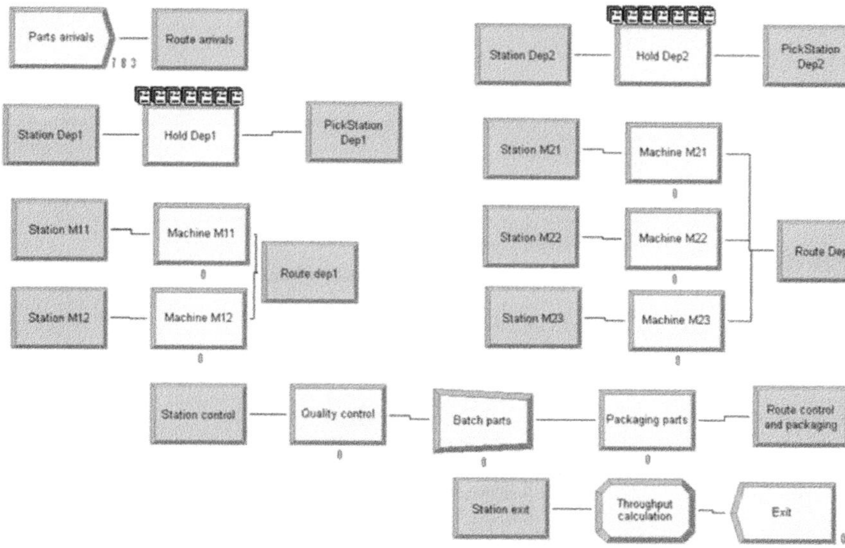

Figure 6.45. *Simulation model with control and packaging. For a color version of the figure, see www.iste.co.uk/jerbi/discrete.zip*

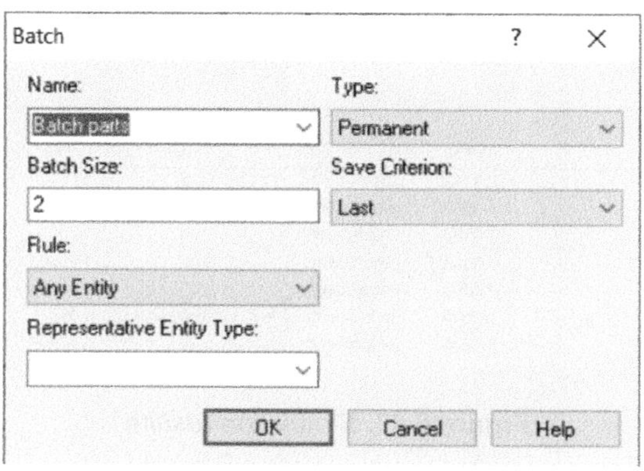

Figure 6.46. *Configuration of the "Batch" module*

The simulation results are as follows (Figure 6.47).

Number In	Value
Entity 1	788.00

Number Out	Value
Entity 1	339.00

WIP	Average	Half Width	Minimum Value	Maximum Value
Entity 1	281.62	(Correlated)	55.0000	507.00

Queue

Time

Waiting Time	Average	Half Width	Minimum Value	Maximum Value
Batch parts.Queue	2.0055	(Insufficient)	0.00	6.0041
Hold Dep1.Queue	182.97	(Correlated)	34.2995	332.56
Hold Dep2.Queue	270.33	(Insufficient)	48.0000	492.00
Packaging parts.Queue	0.00	(Insufficient)	0.00	0.00
quality control.Queue	2.1517	(Insufficient)	0.00	3.9908

Number Waiting	Average	Half Width	Minimum Value	Maximum Value
Batch parts.Queue	0.5017	(Insufficient)	0.00	2.0000
Hold Dep1.Queue	138.01	(Correlated)	23.0000	250.00
Hold Dep2.Queue	136.33	(Insufficient)	24.0000	251.00
Packaging parts.Queue	0.00	(Insufficient)	0.00	0.00
quality control.Queue	0.5403	(Insufficient)	0.00	1.0000

Resource

Usage

Instantaneous Utilization	Average	Half Width	Minimum Value	Maximum Value
control	0.8742	(Insufficient)	0.00	1.0000
M11	1.0000	(Insufficient)	0.00	1.0000
M12	1.0000	(Insufficient)	0.00	1.0000
M21	1.0000	(Insufficient)	0.00	1.0000
M22	1.0000	(Insufficient)	0.00	1.0000
M23	1.0000	(Insufficient)	0.00	1.0000
packaging	0.1201	(Insufficient)	0.00	1.0000

Figure 6.47. *Simulation results*

6.5.4. *Packaging and boxes (Match)*

The packaging operation now requires the presence of boxes. These boxes arrive at the packing workstation according to an exponential distribution with an average of 8 minutes. This is the role of the "Create" module "box arrivals".

To match the boxes to the two grouped products packages, a "Match" module is used. This module has two queues: one for the packages and another for the boxes.

By specifying the number “2” in the “Number to Match” box, this module is required to match two entities, each belonging to one of the queues. If one of these entities is not available, the other must wait. Once matched, the two entities are grouped by a “Batch” module before moving on to the packaging operation (Figures 6.48 and 6.49).

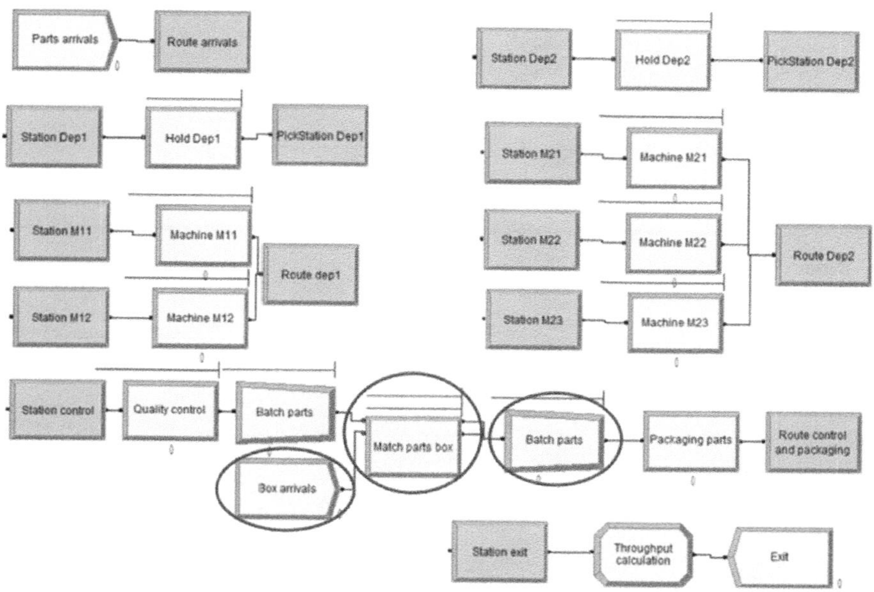

Figure 6.48. *Simulation model with control and packaging, addition of boxes. For a color version of the figure, see www.iste.co.uk/jerbi/discrete.zip*

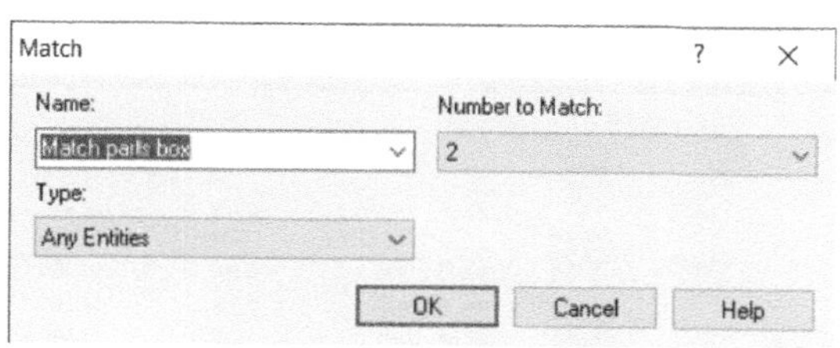

Figure 6.49. *Configuration of the “Match” module*

The new simulation results are shown in Figure 6.50.

Number In	Value			
Entity 1	782.00			
Entity 2	247.00			

Number Out	Value
Entity 1	336.00
Entity 2	224.00

WIP	Average	Half Width	Minimum Value	Maximum Value
Entity 1	278.98	(Correlated)	58.0000	504.00
Entity 2	11.8614	(Correlated)	0.00	25.0000

Queue

Time

Waiting Time	Average	Half Width	Minimum Value	Maximum Value
Batch parts box.Queue	0.00	(Insufficient)	0.00	0.00
Batch parts.Queue	2.0316	(Insufficient)	0.00	6.2014
Hold Dep1.Queue	179.35	(Correlated)	32.7789	325.68
Hold Dep2.Queue	270.33	(Insufficient)	48.0000	492.00
Match parts box.Queue1	2.4869	(Insufficient)	0.00	30.4818
Match parts box.Queue2	77.0686	(Insufficient)	0.00	188.40
Packaging parts.Queue	0.04450874	(Insufficient)	0.00	2.1805
quality control.Queue	2.1352	(Insufficient)	0.00	3.9908

Number Waiting	Average	Half Width	Minimum Value	Maximum Value
Batch parts box.Queue	0.00	(Insufficient)	0.00	2.0000
Batch parts.Queue	0.5086	(Insufficient)	0.00	2.0000
Hold Dep1.Queue	134.36	(Correlated)	26.0000	247.00
Hold Dep2.Queue	136.33	(Insufficient)	24.0000	251.00
Match parts box.Queue1	0.3122	(Insufficient)	0.00	4.0000
Match parts box.Queue2	11.5737	(Insufficient)	0.00	24.0000
Packaging parts.Queue	0.00558832	(Insufficient)	0.00	1.0000
quality control.Queue	0.5362	(Insufficient)	0.00	1.0000

Resource

Usage

Instantaneous Utilization	Average	Half Width	Minimum Value	Maximum Value
control	0.8703	(Insufficient)	0.00	1.0000
M11	1.0000	(Insufficient)	0.00	1.0000
M12	1.0000	(Insufficient)	0.00	1.0000
M21	1.0000	(Insufficient)	0.00	1.0000
M22	1.0000	(Insufficient)	0.00	1.0000
M23	1.0000	(Insufficient)	0.00	1.0000
packaging	0.1382	(Insufficient)	0.00	1.0000

Figure 6.50. *Simulation results second version*

6.6. Resource group, schedule and record results (Set, Schedule, Record)

6.6.1. *Description and data*

The aim of this manipulation is to introduce the use of the resource "Set", the "Record" module and the "Schedule" resource capacity calendars.

A vehicle technical control service agrees to carry out controls on light and heavy vehicles. This service is made up of two reception workstations, depending on the type of vehicle, a control workstation and a reporting workstation. The two types of vehicles arrive according to the parameters described in Table 6.3.

Type of vehicle	Inter-arrival times (minutes)	Batch size
Light	Expo (9)	1
Heavy	Expo (18)	2

Table 6.3. *Arrivals of vehicles*

The vehicle processing times for the different technical control workstations are given in Table 6.4.

Type of vehicle	Reception workstation	Control workstation	Reporting workstation
Light	N(3.1)	T (9,12,13)	UNIF [2, 4]
Heavy	N(7.3)	Weib (10, 4.2)	

Table 6.4. *Vehicle processing time (minutes)*

(1) Simulate the model during an 8 hour working day replication and collect the following performance indexes:

– resource utilization rates;

– the average number of vehicles waiting in front of the four workstations;

– the MFT for all vehicle types.

(2) In order to improve the performance of the control workstation, a second operator is called in as reinforcement. The first operator works for 8 hours of the day, while the second operator only works for the first 4 hours:

– modify the simulation model and run a three-day simulation;

– plot the control workstation resource utilization state;

– collect the following performance indexes: the average number of vehicles waiting in the queues of the four workstations and the MFT for all types of vehicles.

6.6.2. *Simulation model*

The control times of the two types of vehicles adhere to distinct statistical distributions. Consequently, these times will be assigned to the entities representing these vehicles as an attribute labeled "CT". Subsequently, this attribute will be used to indicate the processing time in the "Process" named "Control workstation" (Figures 6.51–6.53).

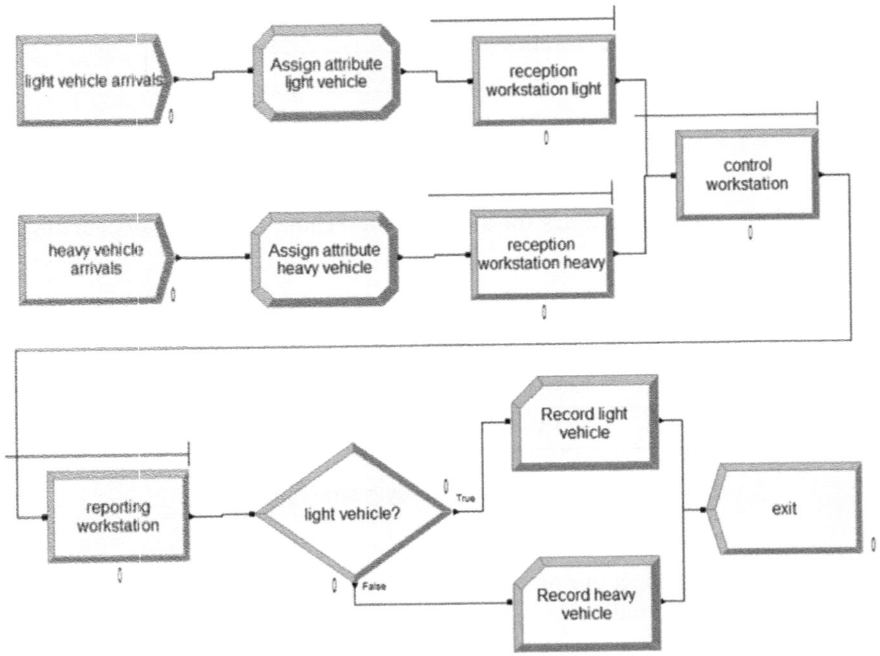

Figure 6.51. *Simulation model*

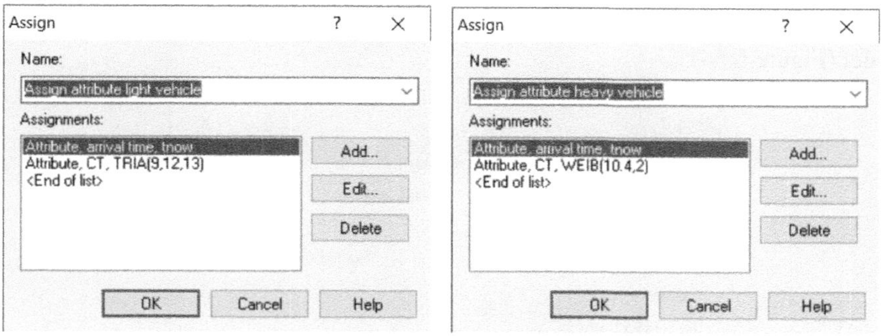

Figure 6.52. *Configuration of the "Assign" modules*

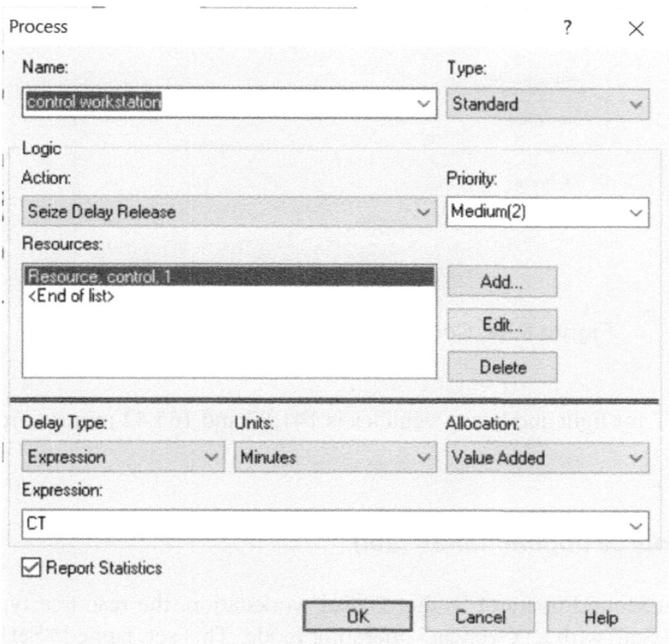

Figure 6.53. *Configuration of the "Process" module "Control workstation"*

To gather the MFT for the two types of vehicles, two "Record" modules are positioned at the end of the model, just before the "Dispose" module. These modules record the interval between the time of the entity's entry into the system and its time of passage through these modules. The "Record" was set as "Time Interval" and a "time entry" attribute was specified as the origin of the interval. This attribute was previously assigned to the entities through the “Assign” modules at the start of the model (Figure 6.54).

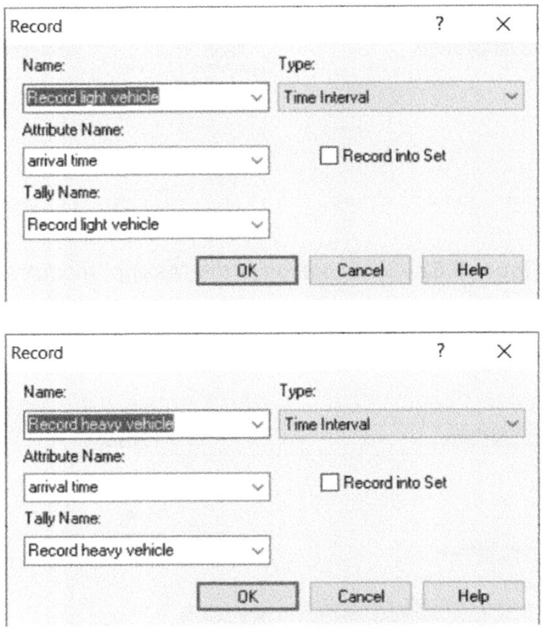

Figure 6.54. *Configuration of the “Record” modules*

The MFT for light and heavy vehicles is 141.87 and 165.42 minutes, respectively (Figure 6.55).

6.6.3. *Resource augmentation (set)*

To add a second operator to the control workstation, the resource type must be changed to “Set” with a “Cyclical” operating mode. This set, named “Set1”, is made up of two operators, “control 1” and “control 2”, which are assigned to it through the “Set-Basic Process” module (Figures 6.56 and 6.57).

The two control operators do not work during the same periods. In fact, the first operator works for 8 hours of the day, while the second operator only works for the first 4 hours. To assign working hours to each operator, the “Resource Basic Process” module is used. This involves changing the “Type” of resources “control 1” and “control 2” from the “Fixed Capacity” type to a “Schedule” type. It is worth noting that different calendar names must be assigned to the two resources (Figure 6.58).

Number In

	Value
light	46.0000
heavy	58.0000

Number Out

	Value
light	19.0000
heavy	27.0000

WIP

	Average	Half Width	Minimum Value	Maximum Value
light	12.1584	(Insufficient)	0.00	27.0000
heavy	19.1125	(Insufficient)	0.00	33.0000

Interval

	Average	Half Width	Minimum Value	Maximum Value
Record light vehicle	141.87	(Insufficient)	16.8482	249.11
Record heavy vehicle	165.42	(Insufficient)	22.0859	267.21

Wait Time

	Average	Half Width	Minimum Value	Maximum Value
light	124.11	(Insufficient)	0.00	232.82
heavy	145.79	(Insufficient)	9.9240	249.19

Number Waiting

	Average	Half Width	Minimum Value	Maximum Value
Poste de control technique.Queue	26.9765	(Insufficient)	0.00	55.0000
Poste emission rapport.Queue	0.00	(Insufficient)	0.00	0.00
Poste reception leger.Queue	0.05316333	(Insufficient)	0.00	2.0000
Poste reception lourd.Queue	1.8143	(Insufficient)	0.00	6.0000

Instantaneous Utilization

	Average	Half Width	Minimum Value	Maximum Value
control	0.9952	(Insufficient)	0.00	1.0000
rapport	0.2948	(Insufficient)	0.00	1.0000
reception leger	0.2769	(Insufficient)	0.00	1.0000
reception lourd	0.8599	(Insufficient)	0.00	1.0000

Figure 6.55. *Simulation results*

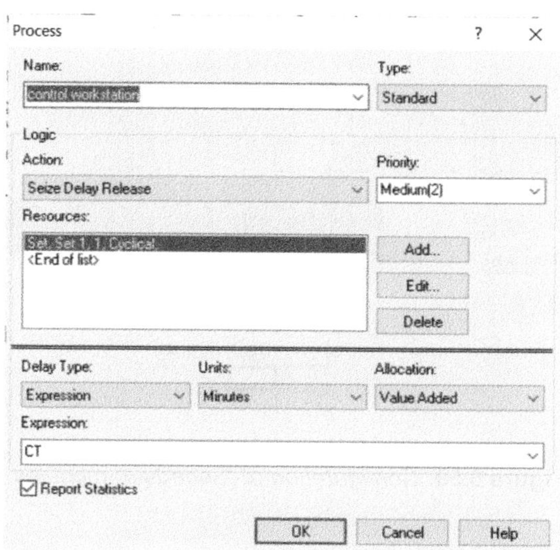

Figure 6.56. *Configuration of the “Process” module “Control workstation”*

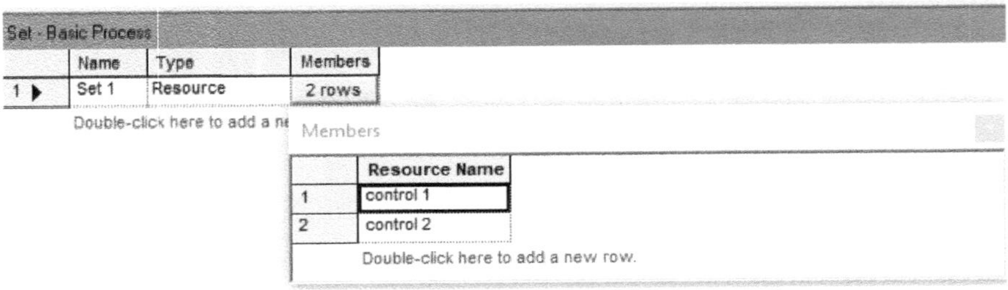

Set - Basic Process

	Name	Type	Members
1 ▶	Set 1	Resource	2 rows

Double-click here to add a ne

Members

	Resource Name
1	control 1
2	control 2

Double-click here to add a new row.

Figure 6.57. *Configuration of the "Set" module*

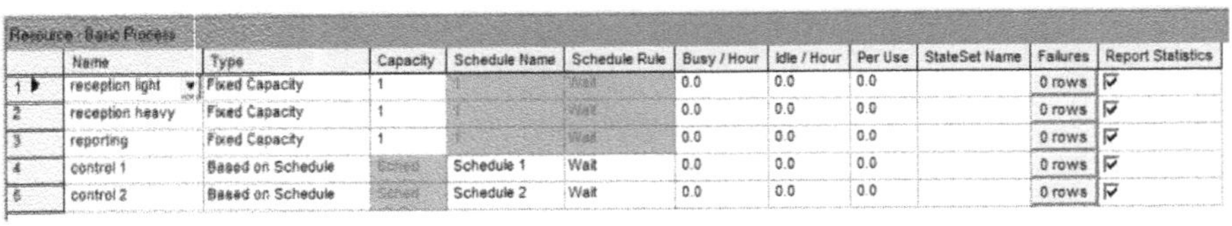

Resource - Basic Process

	Name	Type	Capacity	Schedule Name	Schedule Rule	Busy / Hour	Idle / Hour	Per Use	StateSet Name	Failures	Report Statistics
1 ▶	reception light	Fixed Capacity	1	1	Wait	0.0	0.0	0.0		0 rows	☑
2	reception heavy	Fixed Capacity	1	1	Wait	0.0	0.0	0.0		0 rows	☑
3	reporting	Fixed Capacity	1	1	Wait	0.0	0.0	0.0		0 rows	☑
4	control 1	Based on Schedule	Sched	Schedule 1	Wait	0.0	0.0	0.0		0 rows	☑
5	control 2	Based on Schedule	Sched	Schedule 2	Wait	0.0	0.0	0.0		0 rows	☑

Figure 6.58. *Configuration of "Resource" modules*

The schedules parameters are to be specified through the "Schedule-Basic Process" module.

The "schedule 1" parameter has only one 8 hour interval with a capacity of one operator. The "schedule 2" parameter has two intervals of 4 hours each with a capacity of one operator for the first 4 hours and no operator for the other 4 hours (Figure 6.59).

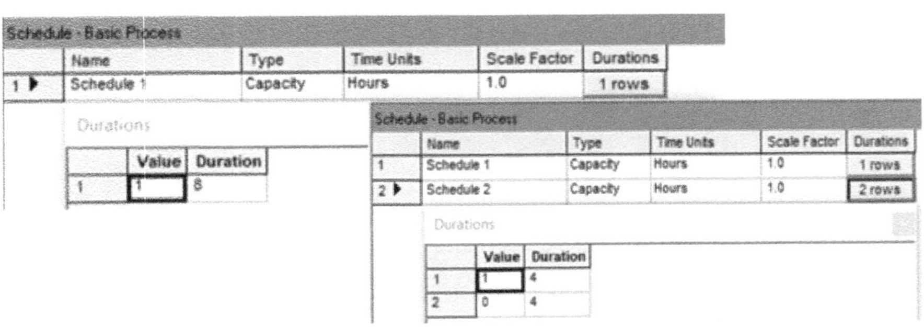

Schedule - Basic Process

	Name	Type	Time Units	Scale Factor	Durations
1 ▶	Schedule 1	Capacity	Hours	1.0	1 rows

Durations

	Value	Duration
1	1	8

Schedule - Basic Process

	Name	Type	Time Units	Scale Factor	Durations
1	Schedule 1	Capacity	Hours	1.0	1 rows
2 ▶	Schedule 2	Capacity	Hours	1.0	2 rows

Durations

	Value	Duration
1	1	4
2	0	4

Figure 6.59. *Configuration of "Schedule" modules*

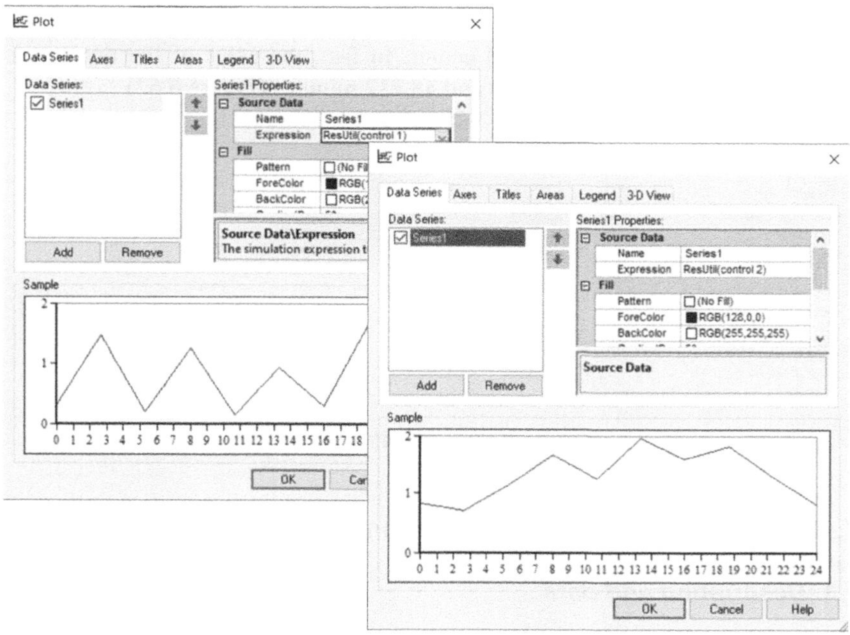

Figure 6.60. *Setting the "PLOT" icons to plot the use of control station resources*

To plot the state graphs for the use of the resources of the control workstation, the "PLOT" icon is used, specifying the expression. These expressions are respectively "ResUtil(control 1)" and "ResUtil(control 2)" (Figure 6.60).

We verify that the first operator's shift lasts for 8 hours and that the second operator's shift spans 4 hours, with a subsequent 4 hour break (Figure 6.61).

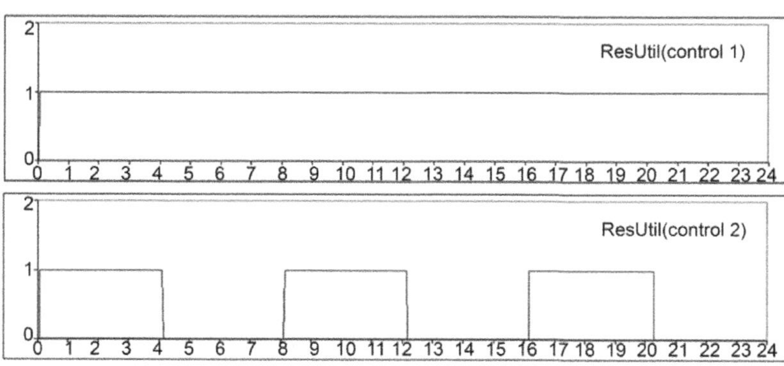

Figure 6.61. *Control workstation resource utilization graphs*

The addition of one operator to the control workstation, even for a half working day, completely unclogs the control center. In fact, the MFT of the two types of vehicles decreases to 42.200 minutes and 48.212 minutes (Figure 6.62).

Number Waiting	Average	Half Width	Minimum Value	Maximum Value
control workstation.Queue	6.4019	(Insufficient)	0.00	22.0000
reception workstation heavy.Queue	0.6586	(Insufficient)	0.00	5.0000
reception workstation light.Queue	0.0936	(Insufficient)	0.00	2.0000
reporting workstation.Queue	0.06737104	(Insufficient)	0.00	2.0000

Interval	Average	Half Width	Minimum Value	Maximum Value
Record heavy vehicule	48.2125	(Insufficient)	12.0668	155.37
Record light vehicule	42.2000	(Insufficient)	15.4311	139.66

Figure 6.62. *Simulation results*

6.7. Supply chain (Request, Transport and Free)

6.7.1. *Description and data*

The aim of this manipulation is to introduce the use of the "Request", "Transport" and "Free" modules in the context of modeling a supply chain.

In a supply chain composed of two companies and a single customer, the raw materials arrive at the two companies in batches of two according to the inter-arrival times, which respectively follow the statistical distributions EXP(3) hours and UNIF [1, 2] hours. These companies process these raw materials for two types of product according to manufacturing times which respectively follow the distributions T(1;2;4) hours and T(1;1.5;2) hours.

Once manufactured, the products are put on hold in stores that are specific to each company. These stores wait for the customer to send an order to release the corresponding products. The parameters for sending orders by the customer are summarized in Table 6.5.

<table>
<tr><th>Business</th><th>Inter-order time</th><th>Number of products ordered</th></tr>
<tr><td>1</td><td rowspan="2">NORM (3.1) h</td><td>2</td></tr>
<tr><td>2</td><td>2</td></tr>
</table>

Table 6.5. *Order Settings*

The products from each order are grouped and then sent to the customer by one of the two trucks from the fleet shared by the two companies. The distance between companies and the customer is 100 km. In addition, any loaded truck travels at a speed of 50 km/h, while during the return trip empty it can travel at 90 km/h speed.

(1) Create the model of this supply chain, simulate this model for a replication length of 100 days at a rate of 24 working hours per day and collect the relevant performance indexes.

(2) What would happen if each company only uses its own truck for its deliveries without any sharing with the other company?

6.7.2. *Simulation model*

The simulation model includes two types of flows: a physical flow of products and an informational flow of orders. The two "Create" modules "raw material 1" and "raw material 2" create the entities representing the physical flow. The "Create" module "customer order" creates the entities representing the information flow.

The companies are modeled by two "Process" modules, followed by two "Hold" modules that materialize the respective stores. Once an order entity is generated, it passes through two "Signal" modules that notify the two stores of the arrival of new orders so that they release two products each. The four products are subsequently grouped by a "Batch" module, called "product grouping for order" (Figures 6.63–6.65).

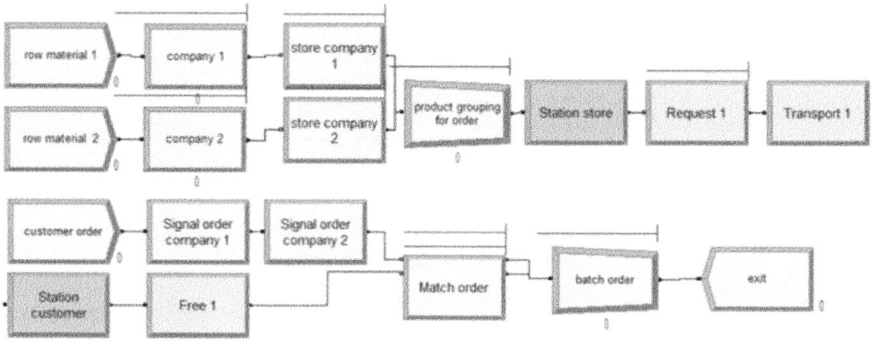

Figure 6.63. *Simulation model. For a color version of the figure, see www.iste.co.uk/jerbi/discrete.zip*

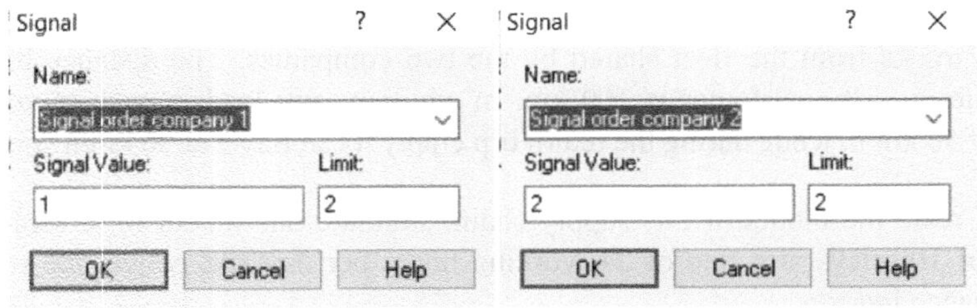

Figure 6.64. *Configuration of "Signal" modules*

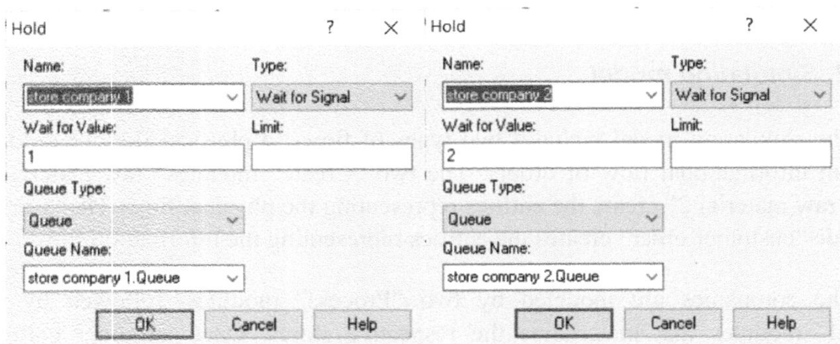

Figure 6.65. *Configuration of the "Hold" modules*

The entire quantity of the order is sent to a "Request" module that puts it on hold until one of the trucks is free. In this case, this module calls it back to the "Station" named "Store station" with a speed of 90 km/h (Figure 6.66).

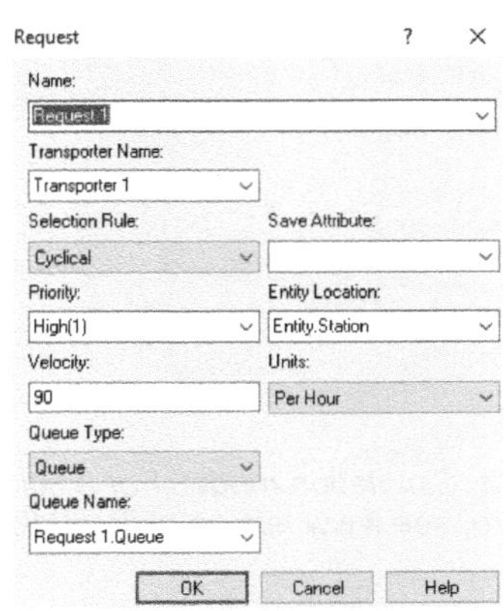

Figure 6.66. *Configuration of the "Request" module*

Once the empty truck arrives at the "Station store", it is loaded with the waiting products. Truck and products then pass through the "Transport" module, which sends them to the corresponding station, "Station customer", with a speed of 50 km/h (Figure 6.67). When the truck arrives at the customer, it is unloaded by the "Free" module (Figure 6.68).

The number of trucks in the "Transporter 1" fleet is specified in the "Transporter – advanced Transfer" module (Figure 6.69).

The distance between the "Station store" and the "Station customer" must be specified in the "Distance – advanced transfer" module (Figure 6.70).

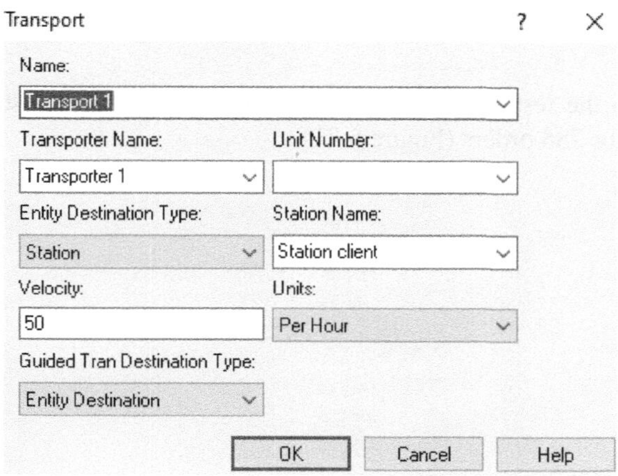

Figure 6.67. *Configuration of the "Transport" module*

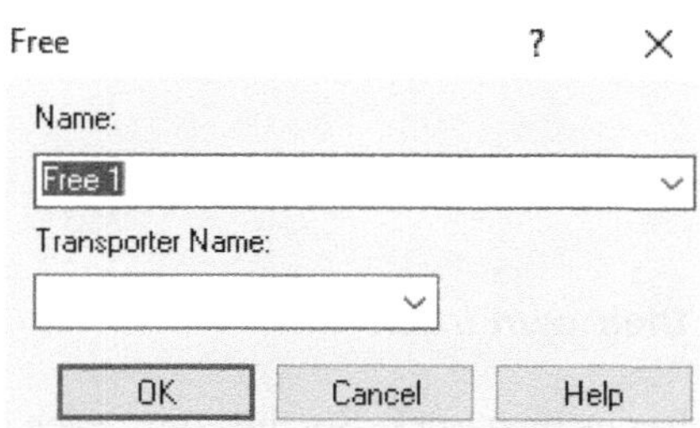

Figure 6.68. *Configuration of the "Free" module*

Transporter - Advanced Transfer

	Name	Number of Units	Type	Distance Set	Velocity	Units	Initial Position Status	Report Statistics
1	Transporter 1	2	Free Path	Transporter 1.Distance	1	Per Hour	2 rows	☑

Figure 6.69. *Configuration of the number of trucks in the fleet*

Distance - Advanced Transfer

	Name	Stations
1	Transporter 1.Distance	1 rows

Stations

	Beginning Station	Ending Station	Distance
1	Station store	Station customer	100

Figure 6.70. *Configuration of the supply chain distances*

According to the results, the truck utilization rate level is almost 42% with 651 orders fulfilled for 785 orders (Figure 6.71).

Number In

	Value
Entity company1	2127.00
Entity company2	4004.00
Entity order	785.00

Number Out

	Value
Entity company1	1545.00
Entity company2	2361.00
Entity order	651.00

Instantaneous Utilization

	Average	Half Width	Minimum Value	Maximum Value
Resource company 1	1.0000	(Insufficient)	0.00	1.0000
Resource company 2	1.0000	(Insufficient)	0.00	1.0000

Utilization

	Average	Half Width	Minimum Value	Maximum Value
Transporter 1	0.4219	0,011035956	0.00	1.0000

Figure 6.71. *Simulation results*

6.7.3. *Everyone has their own truck*

Each company now has its own truck. So, they must each call their truck through their own "Request" and "Transport" modules. The two "Transport" modules send the trucks to the customer station that corresponds to their company. It is essential to assign the corresponding transporter to each "Request" (Figures 6.72 and 6.73).

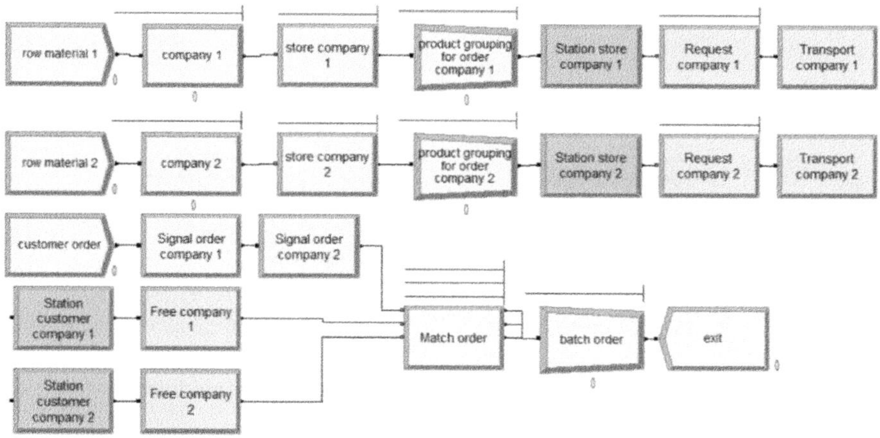

Figure 6.72. *New simulation model. For a color version of the figure, see www.iste.co.uk/jerbi/discrete.zip*

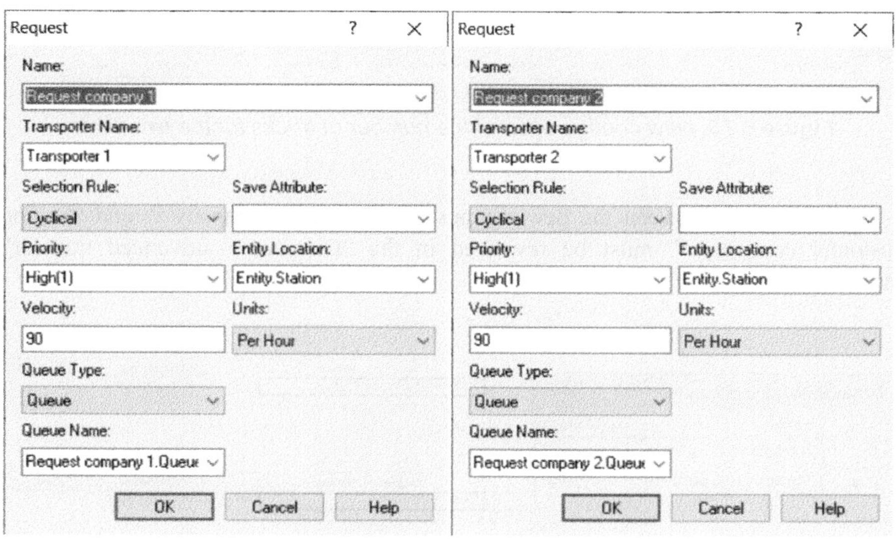

Figure 6.73. *Configuration of the new "Request" modules*

The destination in the two "Transport" modules must be linked to the corresponding client station (Figure 6.74).

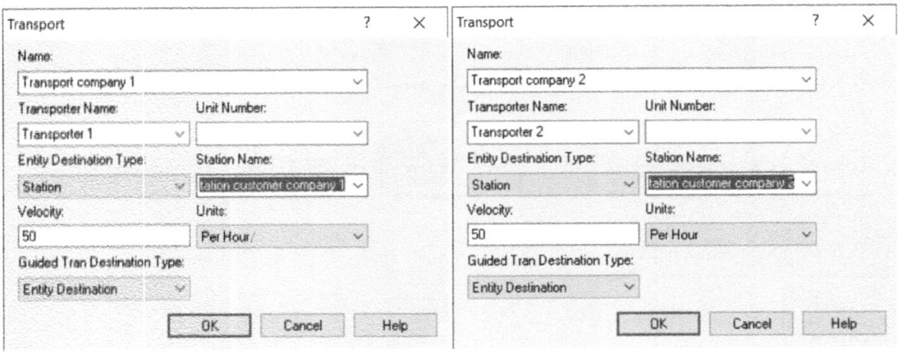

Figure 6.74. *Configuration of the new "Transport" modules*

The number of trucks in the two fleets "Transporter 1" and "Transporter 2" must be updated in the "Transporter – advanced Transfer" module (Figure 6.75).

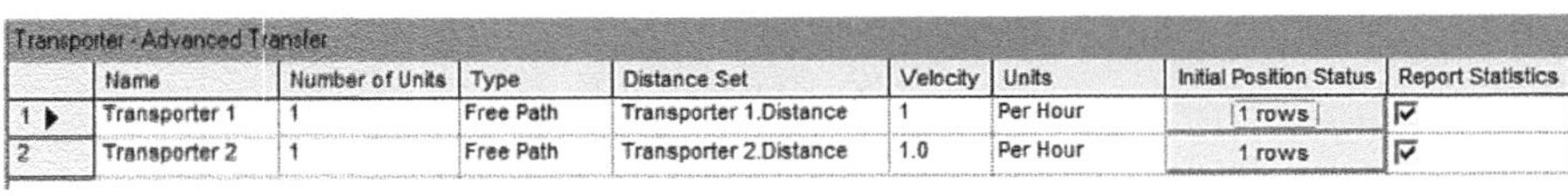

Transporter - Advanced Transfer

	Name	Number of Units	Type	Distance Set	Velocity	Units	Initial Position Status	Report Statistics
1	Transporter 1	1	Free Path	Transporter 1.Distance	1	Per Hour	1 rows	☑
2	Transporter 2	1	Free Path	Transporter 2.Distance	1.0	Per Hour	1 rows	☑

Figure 6.75. *New configuration of the number of trucks for the two fleets*

The distances between the new stations "Station store company i" and "Station customer company i" must be reviewed in the "Distance – advanced transfer" module (Figure 6.76).

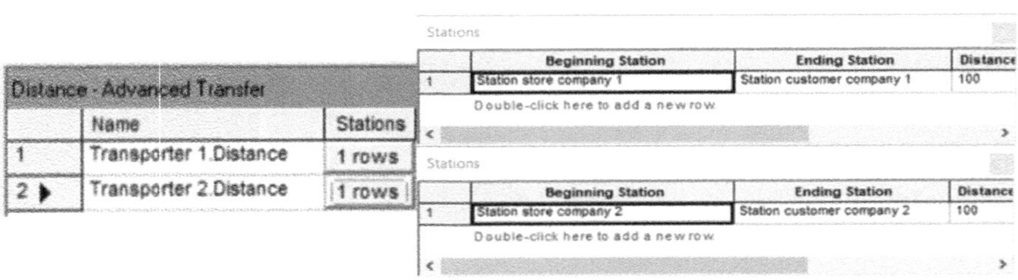

Figure 6.76. *New configuration of supply chain distances*

The new simulation results show that the number of orders honored decreases by 20% to reach 518 orders instead of 651 orders. Yet truck utilization rates increase to 67% and 99%, respectively (Figure 6.77).

Number In	Value			
Entity company1	2137.00			
Entity company2	4512.00			
Entity order	785.00			
Number Out	**Value**			
Entity company1	1554.00			
Entity company2	2072.00			
Entity order	518.00			
Instantaneous Utilization	**Average**	**Half Width**	**Minimum Value**	**Maximum Value**
Resource company 1	1.0000	(Insufficient)	0.00	1.0000
Resource company 2	1.0000	(Insufficient)	0.00	1.0000
Utilization	**Average**	**Half Width**	**Minimum Value**	**Maximum Value**
Transporter 1	0.6723	0,014645836	0.00	1.0000
Transporter 2	0.9913	(Insufficient)	0.00	1.0000

Figure 6.77. *The new simulation results*

7

Simulation within the Manufacturing Sector

7.1. Study of a television assembly line

To remain competitive in both global and local markets, the manufacturing industry hinges significantly on customer satisfaction levels concerning quantity, quality and lead time. Consequently, many companies strive to continuously enhance their performance to meet the challenge of continuous improvement and development of production technologies. It is therefore imperative to identify the parameters influencing manufacturing system performance without incurring additional costs. One viable approach is to use a simulation model of the manufacturing system. This model enables imitating the real dynamic manufacturing system resources and facilitates identifying the main bottlenecks without implementing any change in the real manufacturing system. The main objective of company "X" is to ensure the satisfaction of its two customers, both in terms of product and service. The company operates five workshops dedicated to manufacturing various products. These include the office and household equipment workshop, the energy metro workshop, the electro-optics workshop, the communication equipment workshop and the biomedical workshop. Among these workshops, particular emphasis is placed on the office and household equipment workshop where LCD TVs are assembled. This focus is attributed to the significant demand for LCD TVs, as well as the high labor intensity, large resources and lengthy manufacturing and assembly process of the products compared to others. The factory's main goal is to maximize its daily production output of LCD TVs. The total number of assembly operations required to manufacture these TVs on the existing assembly line is 28.

7.1.1. *Objective*

Drawing from 2018 data, it is evident that the actual average daily production is significantly below the capacity of the actual assembly line. Consequently, this study seeks to enhance the assembly line production output by identifying the causes constraining its performance and efficiency in the assembly line process. Furthermore, it aims to explore strategies to increase throughput while reducing production costs.

7.1.2. *Setting up the TV assembly line*

In the office and household equipment workshop, the television is assembled on an assembly line following successive steps. The list of operations is given in Table 7.1.

Ope	Description	Ope	Description
OP1	Unpack the frame from the cardboard	OP15	Secure the back of the cabinet
OP2	Disassemble the frame	OP16	Load the software
OP3	Paste mat EVA sponge	OP17	AC and DC testing
OP4	Place the screen on the frame	OP18	HDMI and YPBPR testing
OP5	Attach the left and right panels	OP19	AV and coaxial testing
OP6	Place panel sticker, HT cable	OP20	USB and SCART testing
OP7	Attach the EVA mat	OP21	Antenna test
OP8	Attach the boss support	OP22	TV and VGA test
OP9	Attach the power board bracket	OP23	Cleaning
OP10	Attach the power board	OP24	Affix sticker and barcode, power cable packaged
OP11	Fix the main board	OP25	Warranty and manual, remote control, serial number registration
OP12	Fix the terminal up and down	OP26	Screen covering
OP13	Insert the panel and the HV cable	OP27	Install the safety socket
OP14	Insert the speaker and power cable	OP28	Pack in a carton

Table 7.1. *List of television production operations*

7.1.3. *Collection of data*

7.1.3.1. *Processing time for each operation*

The processing time of each operation was delineated as the duration from entering the station to process completion. These durations were meticulously measured in seconds at each workstation using a stopwatch. Table 7.2 summarizes the measurements collected from each operation.

Op	Measurements in seconds
OP1	60 56 58 57 62 64 63 62 68 60 57 62 60 60.5 60 56 58 64 63 62 54 60 68 60 55 54 60 60 60.5
OP2	40 40 42 40 39 38 40 40 37 40 35 35 38 40 40 37 40 43 40 41 42 39 44 40 41 42 43 44 40 42
OP3	65 60 64 65 65 63 64 65 64 65 65 64 65 66 67 68 59 63 62 64 62 63 65 61 60 66 67 66 68 64
OP4	3534 37 35 36 35 38 36 35 34 40 35 39 35 33 36 35 38 36 37 34 37 35 35 35 39 36 40 36 34
OP5	35 36 34 36 34 30 32 35 36 35 35 38 37 34 39 36 37 34 35 35 30 33 36 37 34 34 36 36 34 38
OP6	35 38 36 38 35 37 40 33 35 35 35 38 33 36 32 39 40 39 38 32 36 35 35 35 37 34 38 34 34 35
OP7	70 66 72 70 76 74 69 65 70 72 68 66 75 70 70 70 66 72 70 76 74 69 65 70 72 68 66 75 70 70
OP8	73 71 75 73 74 73 71 72 65 70 76 72 76 76 72 72 747578 68 76 71 72 74 75 78 68 73 74 65
OP9	25 20 23 19 20 24 25 20 23 19 20 23 23 22 25 21 18 24 18 20 23 24 18 23 22 25 21 18 20 24
OP10	40 36 43 34 43 40 40 36 37 42 37 37 45 41 46 42 37 39 40 4546 45 41 39 40 45 43 34 43 40
OP11	60 59 53 62 64 54 68 54 59 58 54 59 64 64 68 70 58 5460 59 53 62 71 56 64 68 70 68 71 56
OP12	80 84 79 85 78 81 77 75 79 84 86 85 80 77 78 80 84 76 85 75 81 75 75 79 84 76 85 80 77 78
OP13	80 76 84 79 83 85 78 81 81 83 82 77 79 80 83 80 76 84 79 83 85 78 81 81 83 82 77 79 80 83
OP14	88 78 88 90 92 84 91 93 87 79 80 91 78 92 80 89 79 89 91 93 85 92 93 88 80 81 92 79 93 81
OP15	90 88 89 97 86 85 89 90 90 104 101 86 90104 101 97 96 99 93 9194 91 97 98 99 97 95 96 92 98
OP16	25 25 29 27 29 25 20 24 20 22 26 20 25 27 25 2324 20 28 29 25 23 24 26 27 26 27 25 23 21
OP17	60 59 58 63 65 62 64 59 56 59 64 59 65 62 62 64 60 63 61 65 62 63 62 60 59 58 63 59 62 58
OP18	50 50 47 48 56 53 52 49 53 58 48 50 56 53 51 50 50 50 47 50 51 50 48 49 50 53 58 48 52 50
OP19	48 45 54 56 4150 45 49 41 39 40 50 48 40 45 39 41 50 39 40 41 39 45 40 50 45 41 49 41 50
OP20	54 48 49 49 57 50 60 60 54 48 49 49 54 56 58 58 60 58 60 49 55 50 49 54 56 58 60 48 60 60
OP21	60 64 57 60 65 56 65 56 59 64 60 65 64 60 58 57 60 65 65 60 65 5759 55 63 59 59 55 63 65
OP22	40 42 35 41 39 38 40 40 39 45 43 46 37 38 42 43 46 37 38 35 44 45 35 44 35 41 39 38 40 39
OP23	6059 63 67 65 57 60 58 64 56 61 66 59 64 61 62 60 63 67 65 57 61 62 58 60 56 59 61 66 59
OP24	40 37 38 38 41 36 37 35 36 38 40 35 40 35 36 35 36 39 39 40 38 37 38 40 41 35 40 35 37 38
OP25	38 36 42 43 36 48 40 45 44 35 40 34 35 40 40 38 37 40 36 42 43 48 45 37 44 35 36 40 34 35
OP26	15 16 12 18 13 15 14 15 18 16 15 16 15 13 14 13 15 16 15 12 18 15 14 18 16 14 13 15 16 15
OP27	17 20 20 14 14 15 16 18 14 15 19 17 12 17 15 20 17 15 18 20 20 14 20 15 18 14 18 17 15 14
OP28	43 40 38 43 36 40 42 42 43 41 37 40 38 39 40 45 39 40 40 42 45 40 40 43 36 42 43 43 41 37

Table 7.2. *Collected measurements of processing times for each operation*

These measurements were subject to analysis with the Input Analyzer tool within Arena software to model them according to statistical distributions. Figures 7.1–7.28 present the modeled expressions of the different assembly operations.

The adequacy of the statistical distributions governing these operations was verified by examining the p-values obtained from statistical tests.

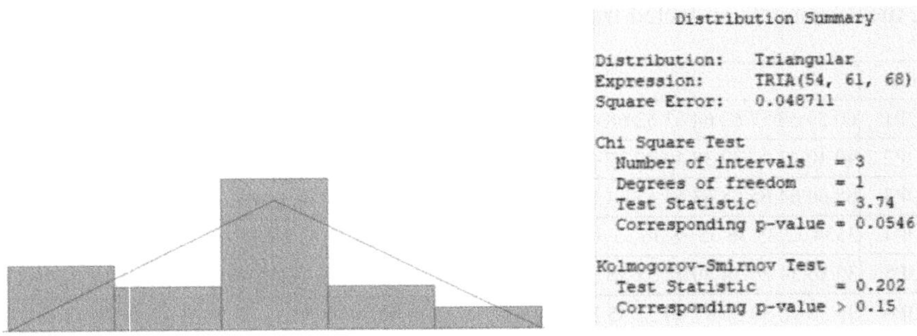

Figure 7.1. *Modeling the processing time of operation OP1*

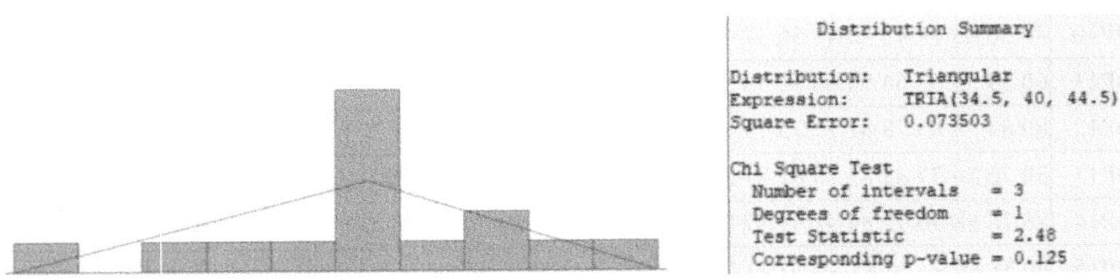

Figure 7.2. *Modeling the processing time of the OP2 operation*

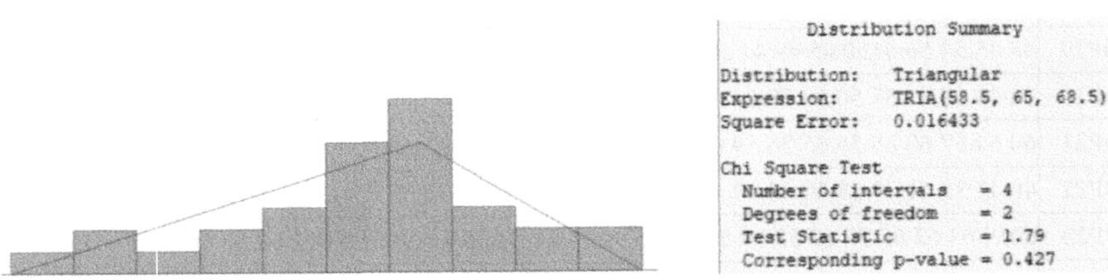

Figure 7.3. *Modeling the processing time of the OP3 operation*

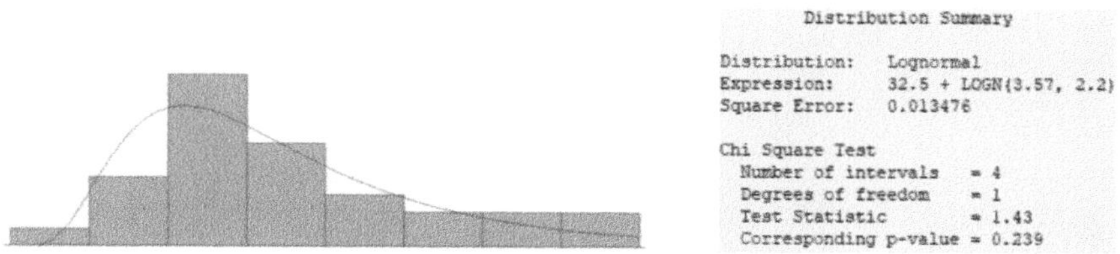

Figure 7.4. *Modeling the processing time of the OP4 operation*

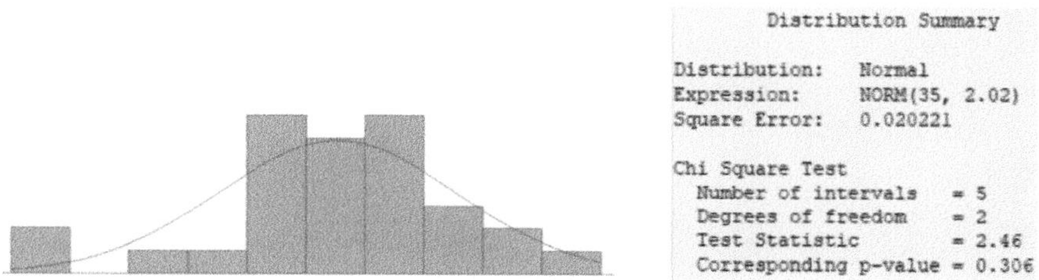

Figure 7.5. *Modeling the processing time of the OP5 operation*

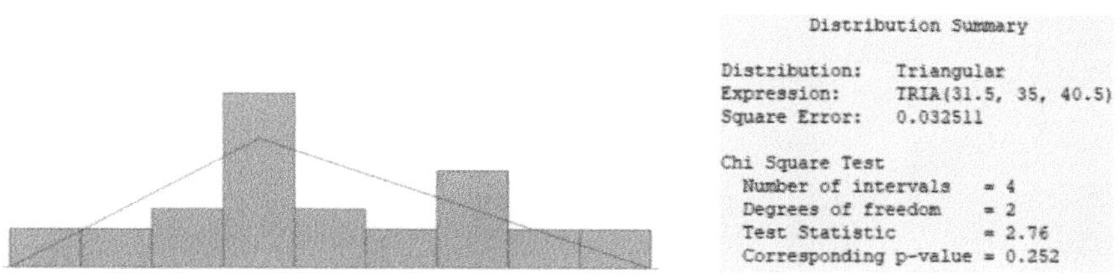

Figure 7.6. *Modeling the processing time of the OP6 operation*

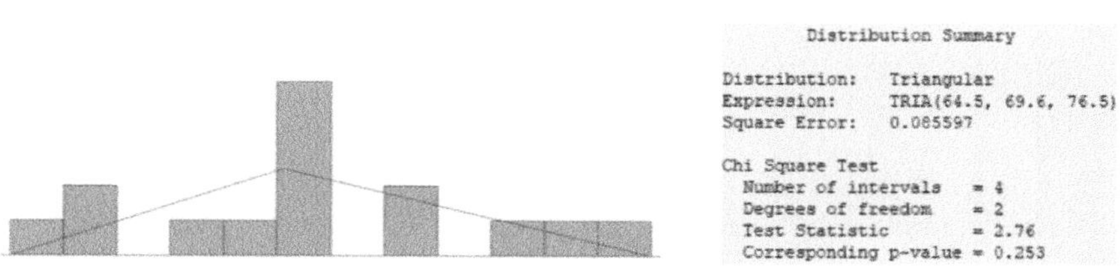

Figure 7.7. *Modeling the processing time of the OP7 operation*

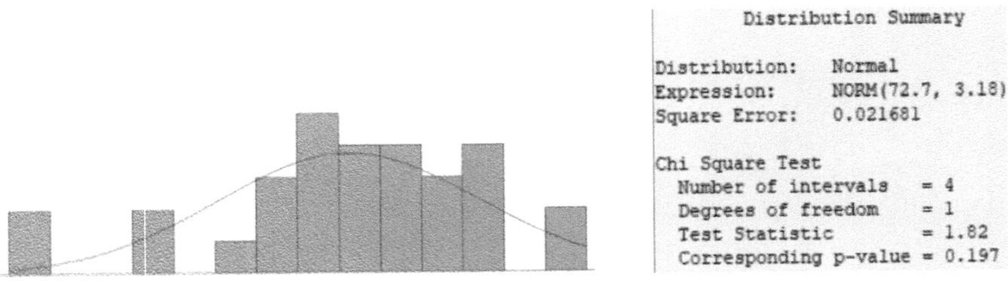

Figure 7.8. *Modeling the processing time of the OP8 operation*

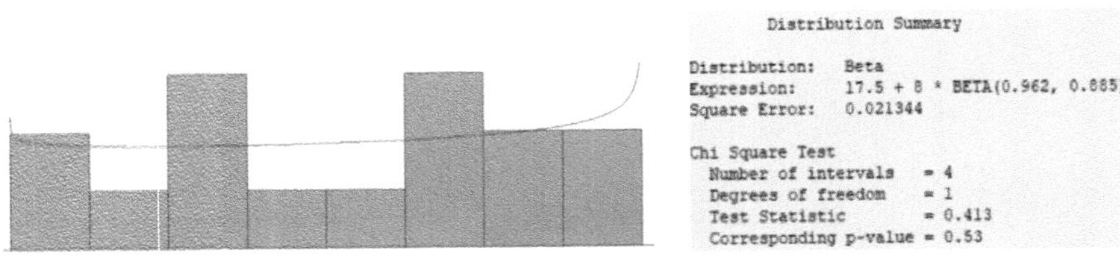

Figure 7.9. *Modeling the processing time of the OP9 operation*

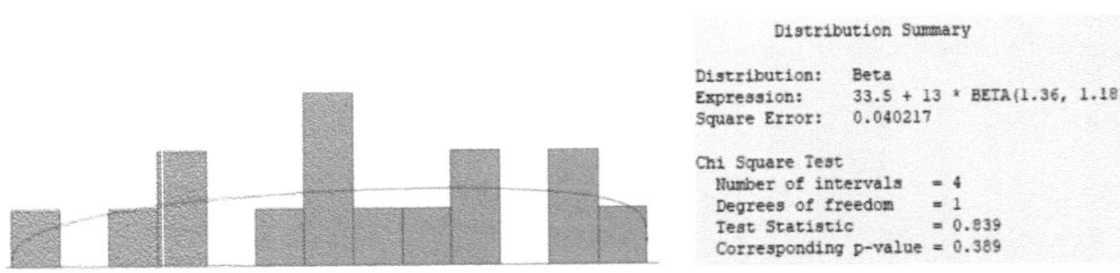

Figure 7.10. *Modeling the processing time of the OP10 operation*

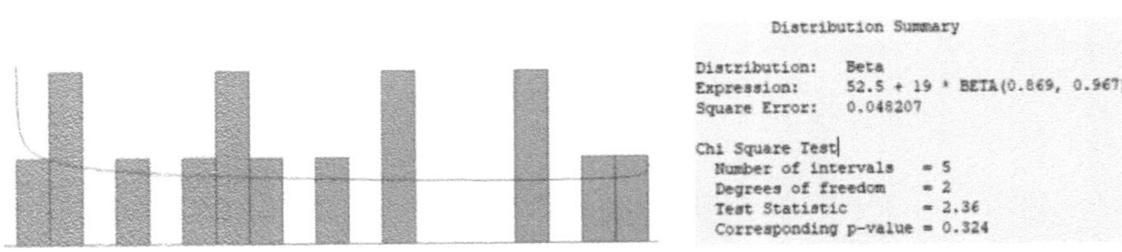

Figure 7.11. *Modeling the processing time of the OP11 operation*

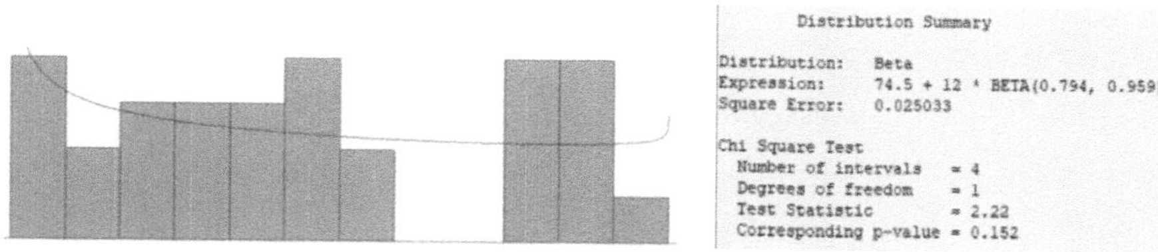

Figure 7.12. *Modeling the processing time of the OP12 operation*

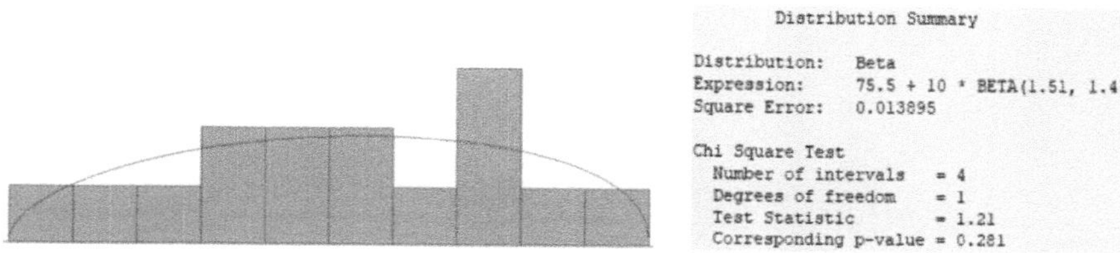

Figure 7.13. *Modeling the processing time of the OP13 operation*

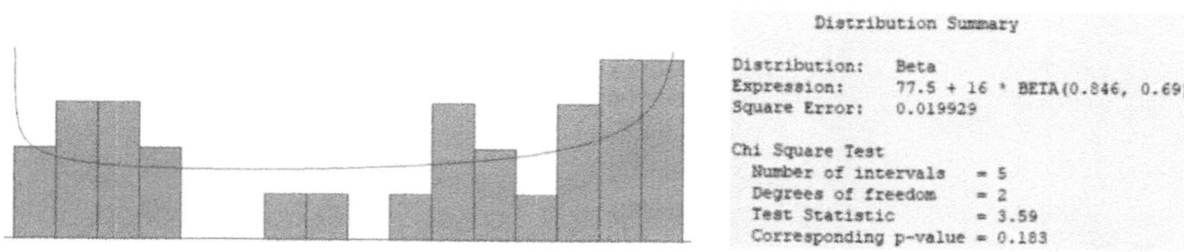

Figure 7.14. *Modeling the processing time of the OP14 operation*

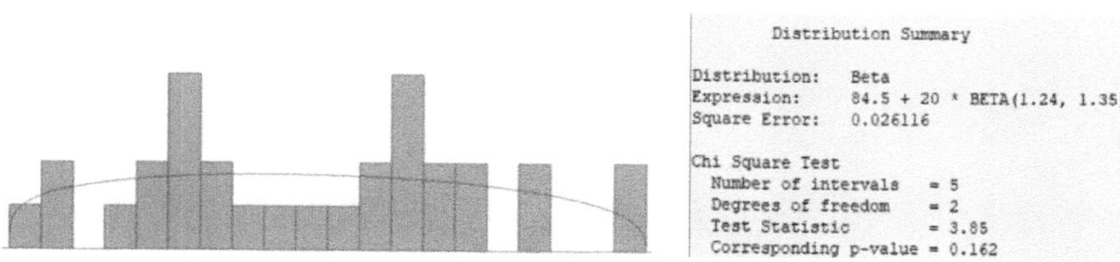

Figure 7.15. *Modeling the processing time of the OP15 operation*

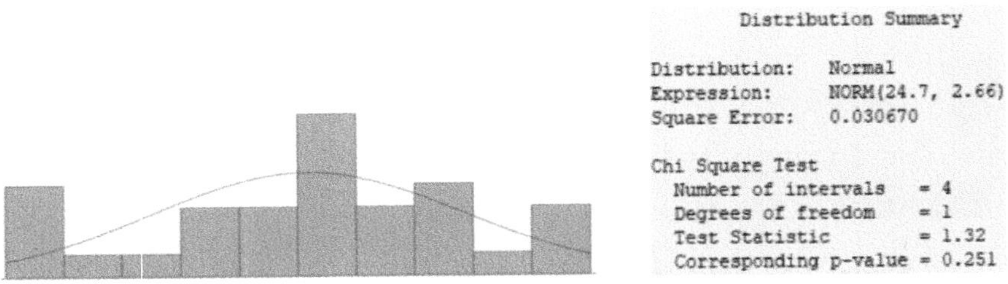

Figure 7.16. *Modeling the processing time of the OP16 operation*

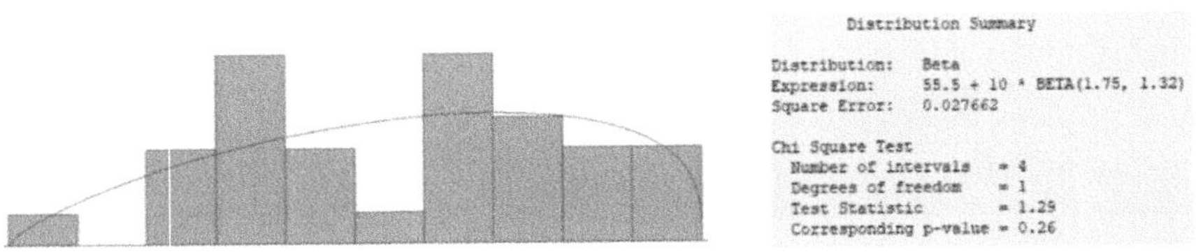

Figure 7.17. *Modeling the processing time of the OP17 operation*

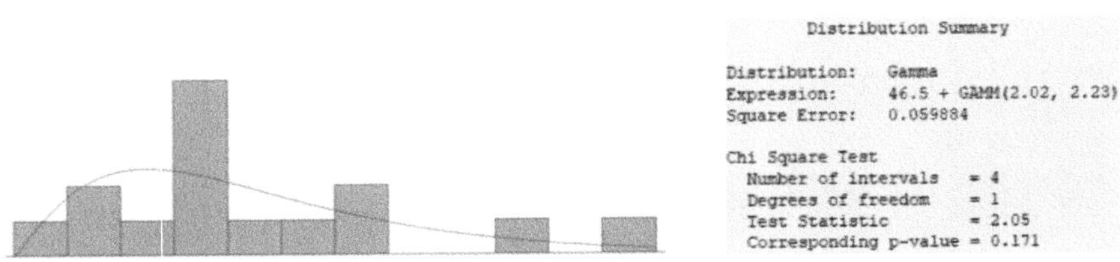

Figure 7.18. *Modeling the processing time of the OP18 operation*

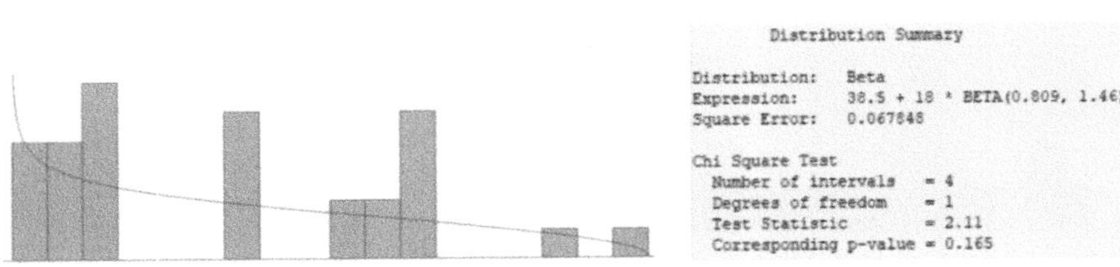

Figure 7.19. *Modeling the processing time of OP19 operation*

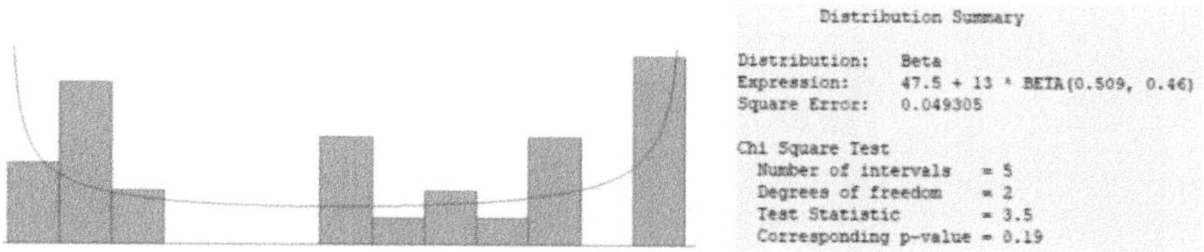

Figure 7.20. *Modeling the processing time of the OP20 operation*

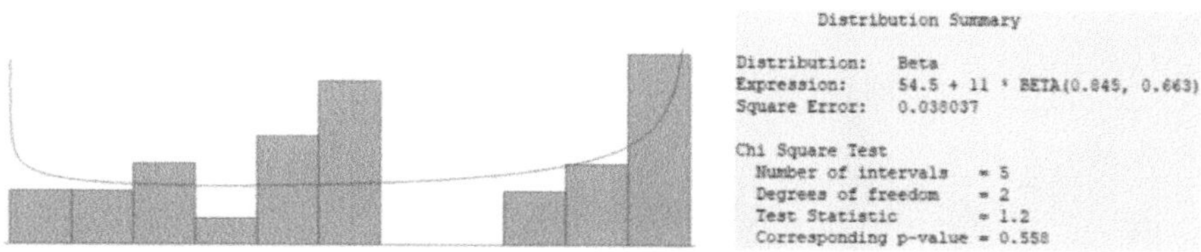

Figure 7.21. *Modeling the processing time of the OP21 operation*

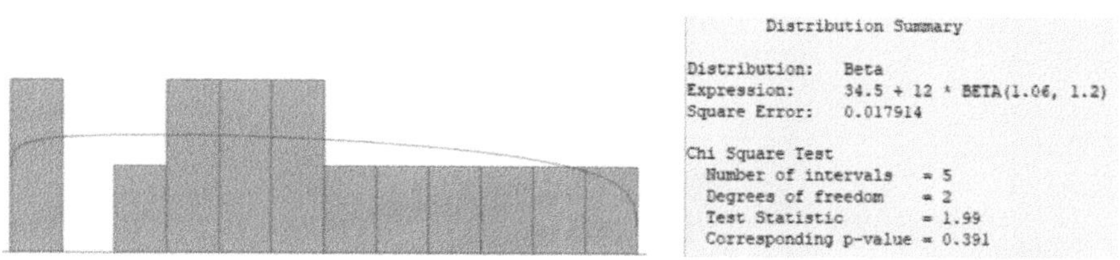

Figure 7.22. *Modeling the processing time of the OP22 operation*

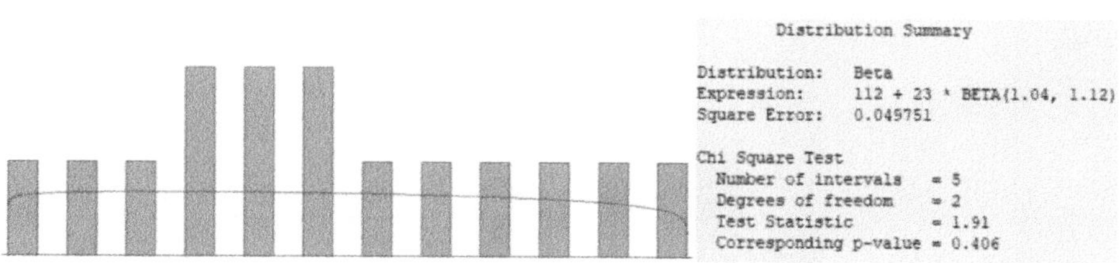

Figure 7.23. *Modeling the processing time of the OP23 operation*

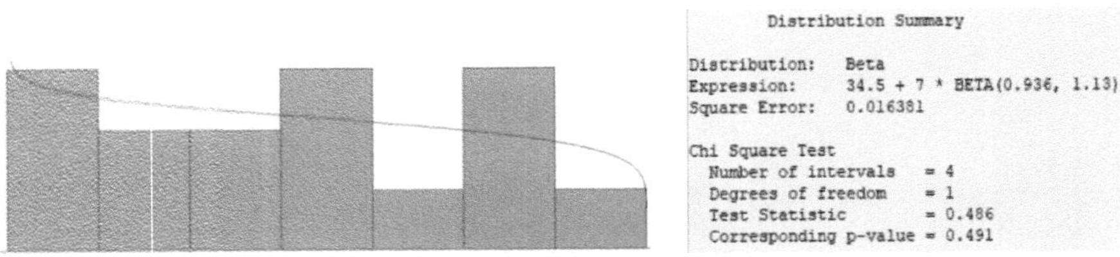

Figure 7.24. *Modeling the processing time of the OP24 operation*

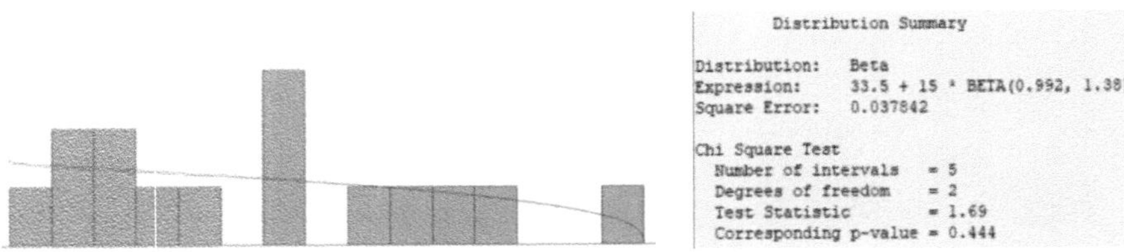

Figure 7.25. *Modeling the processing time of the OP25 operation*

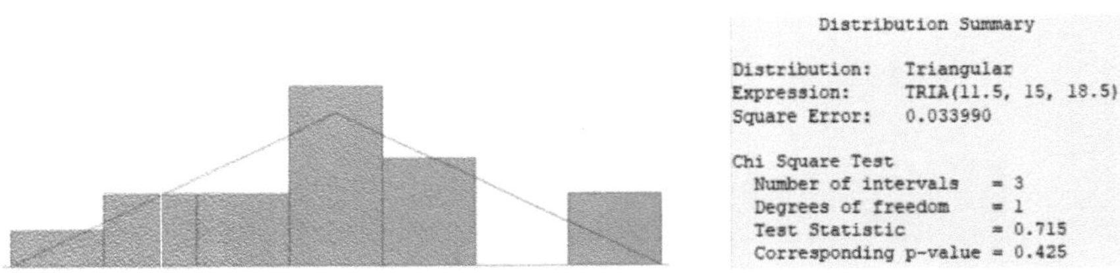

Figure 7.26. *Modeling the processing time of the OP26 operation*

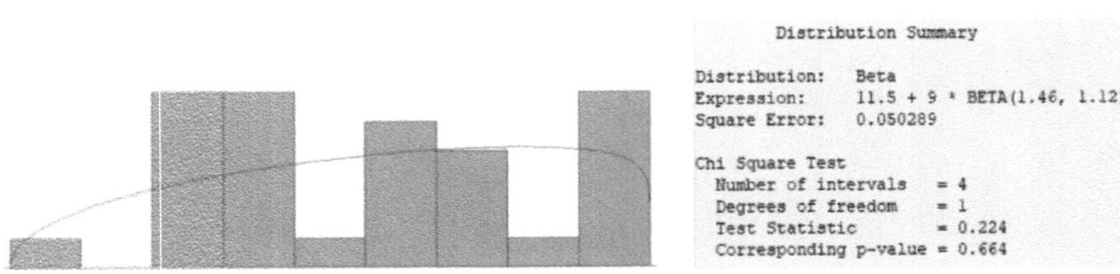

Figure 7.27. *Modeling the processing time of the OP27 operation*

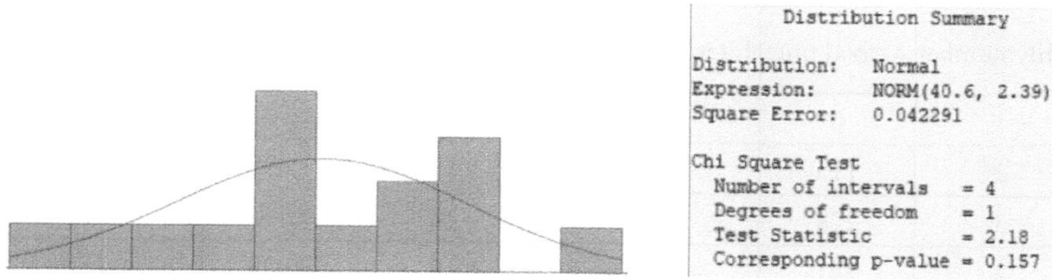

Figure 7.28. *Modeling the processing time of the OP28 operation*

The expressions for the operation times are summarized in Table 7.3.

Op	Expression (in seconds)	Op	Expression (in seconds)
OP1	TRIA (54, 61, 68)	OP15	84.5 + 20 * BETA (1.24, 1.35)
OP2	TRIA (34.5, 40, 44.5)	OP16	NORM (24.7, 2.66)
OP3	TRIA (58.5, 65, 68.5)	OP17	55.5 + 10 * BETA (1.75, 1.32)
OP4	32.5 + LOGN (3.57, 2.2)	OP18	46.5 + GAMM (2.02, 2.23)
OP5	NORM (35, 2.02)	OP19	38.5 + 18 * BETA (0.809, 1.46)
OP6	TRIA (31.5, 35, 40.5)	OP20	47.5 + 13 * BETA (0.509, 0.46)
OP7	TRIA (64.5, 69.6, 76.5)	OP21	54.5 + 11 * BETA (0.845, 0.663)
OP8	NORM (72.7, 3.18)	OP22	34.5 + 12 * BETA (1.06, 1.2)
OP9	17.5 + 8 * BETA (0.962, 0.885)	OP23	112 + 23 * BETA (1.04, 1.12)
OP10	33.5 + 13 * BETA (1.36, 1.18)	OP24	34.5 + 7 * BETA (0.936, 1.13)
OP11	5 2.5 + 19 * BETA (0.869, 0.967)	OP25	33.5 + 15 * BETA (0.992, 1.38)
OP12	74.5 + 12 * BETA (0.794, 0.959)	OP26	TRIA (11.5, 15, 18.5)
OP13	75.5 + 10 * BETA (1.51, 1.4)	OP27	11.5 + 9 * BETA (1.46, 1.12)
OP14	77.5 + 16 * BETA (0.846, 0.69)	OP28	NORM (40.6, 2.39)

Table 7.3. *Expression of operation times*

7.1.3.2. *Inter-arrival times of entities at the assembly line*

The arrival times of several entities on the assembly line were recorded using a stopwatch. Table 7.4 summarizes these records.

Entity number	Arrival time	Inter-arrival times	Entity number	Arrival times	Inter-arrival times
1	0	–	16	1,721	114
2	113	113	17	1,838	117
3	228	115	18	1,953	115
4	341	113	19	2,070	117
5	457	116	20	2,184	114
6	574	117	21	2,300	116
7	686	112	22	2,417	117
8	802	116	23	2,528	111
9	915	113	24	2,640	112
10	1,032	117	25	2,757	117
11	1,146	114	26	2,874	117
12	1,262	116	27	2,989	115
13	1,375	113	28	3,103	114
14	1,490	115	29	3,213	110
15	1,607	117	30	3,330	117

Table 7.4. *Inter-arrival time of entities on the assembly line (in seconds)*

Similar to the processing times of assembly operations, the inter-arrivals time was also modeled using Arena's Input Analyzer tool. This time adhered to a Gamma distribution with a high p-value from the chi-square goodness-of-fit test (Figure 7.29). This underscores the good alignment between the observed measurements and the statistical distribution equation derived.

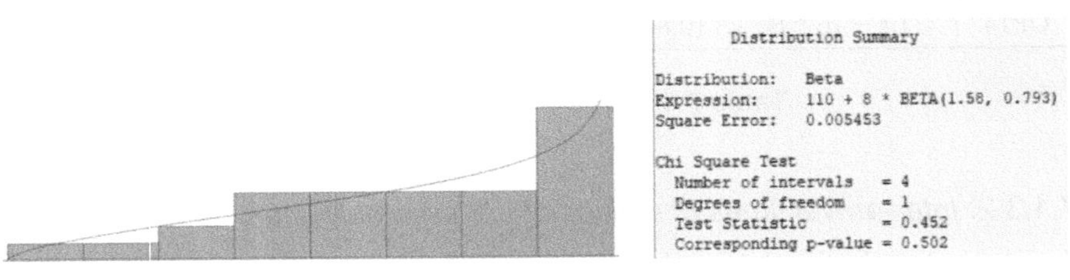

Figure 7.29. *Modeling the inter-arrival time of entities*

7.1.3.3. *Transfer time*

Individual transfers of entities between the different assembly stations are carried out within an estimated constant time in seconds.

7.1.3.4. *Number of operators and operating hours*

The assembly line, made up of 28 stations, designates one operator to each assembly operation. The operational schedule of this assembly line spans 8 hours per day. Conversely, the company schedules assembly line downtime intervals for different reasons such as entry, break, lunchtime and exit/shutdown. Based on historical data and direct observation, actual downtimes were recorded. Table 7.5 provides a summary of the scheduled and actual operating times of the assembly line.

Reason for stopping	Scheduled time (h)	Actual time (h)
Entrance	08:00 a.m.	08:15 a.m.
Break 1	10:00 a.m.–10:15 a.m.	10:00 a.m.–10:30 a.m.
Lunch	12:00 p.m.–1:00 p.m.	12:00 p.m.–1:15 p.m.
Break 2	3:00 p.m.–3:15 p.m.	3:00 p.m.–3:30 p.m.
Exit	5:30 p.m.	5:15 p.m.

Table 7.5. *Assembly line operating schedule*

7.1.4. *Simulation model*

The simulation model, illustrated in the following figure, comprises three distinct sections: one section dedicated to the entry point of the assembly line, a second for the assembly stations and a third for the exit of the televisions (Figure 7.30).

7.1.4.1. *Entry into the assembly line*

The entry section consists of a "Create" module named "Arrivals", which creates entities according to the modeled inter-arrival time (Figure 7.31).

Once created, the entities are routed to a "Route" module, named "Route arrivals", which then transfers them to the first station "OP1" of the assembly line. The transfer time is 4 seconds (Figure 7.32).

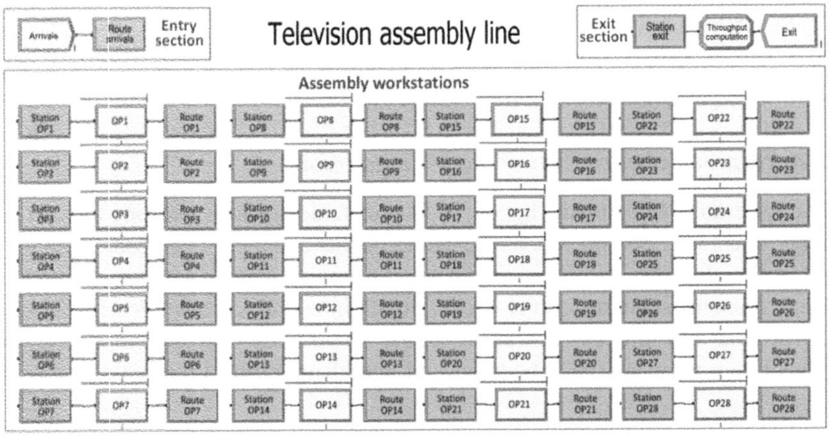

Figure 7.30. *TV assembly line assembly model. For a color version of the figure, see www.iste.co.uk/jerbi/discrete.zip*

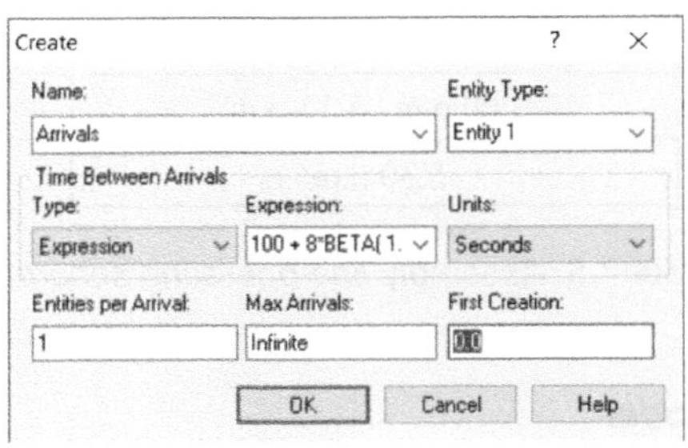

Figure 7.31. *Configuration of the "Create" module entry section, television assembly line*

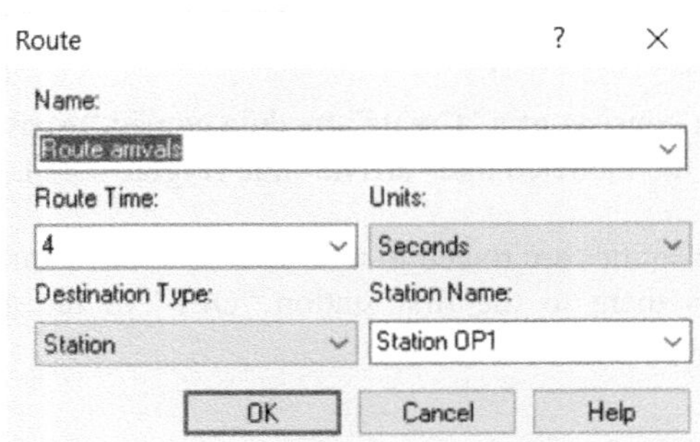

Figure 7.32. *Configuration of the "Route" module entry section, television assembly line*

7.1.4.2. *Assembly workstations*

This section encompasses the 28 assembly stations, with each modeled by a "Process" module preceded by a "Station" module and followed by a "Route" module (Figure 7.33).

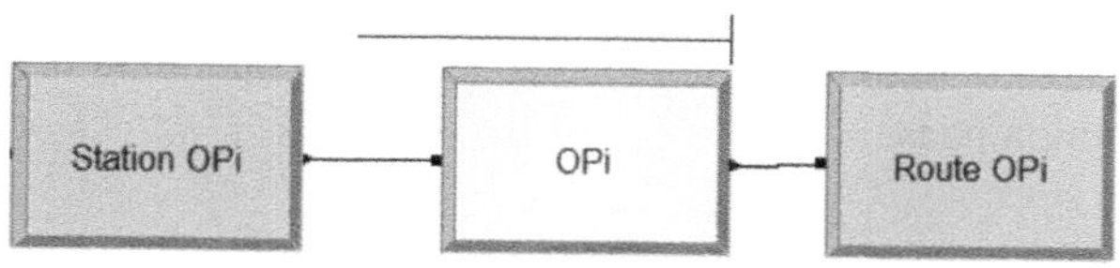

Figure 7.33. *"i" assembly workstation, television assembly line. For a color version of the figure, see www.iste.co.uk/jerbi/discrete.zip*

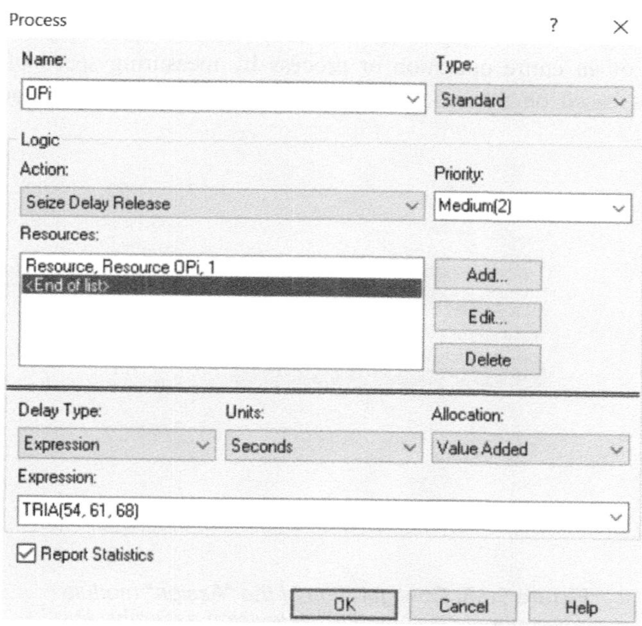

Figure 7.34. *Configuration of the "Process" module of the assembly workstation, television assembly line*

The "Station OPi" module acts as an entry point to the assembly workstation where entities from the previous assembly workstation are directed via the "Route OPi-1" module. The "Route OPi" module therefore transfers the entities processed by station "i" to station "i + 1". The configuration of this module remains identical to that used in Figure 7.32, only the name of the destination station changes.

The “Process” module, labeled “OPi”, operates the “Seize Delay Release” type. It adds value to each entity corresponding to the assembly operation. The configuration of each “Process” follows a process with minor modifications from one “Process” to another (Figure 7.34). These modifications are situated in the “Resources” and “Expression” sections of the “Delay type”. Specifically, a distinct resource is assigned to each “Process” module. Furthermore, each “Process” is assigned a time corresponding to the modeled assembly operation duration gathered in the data collection phase.

7.1.4.3. *Exit point*

In addition to the two “Station” and “Dispose” modules respectively named “Station exit” and “Exit”, this section also includes an “Assign” module. This is used to calculate the instantaneous production rate or throughput. This index is a very widespread concept in the industrial and production context (Addi Ait 2000). It denotes the rate at which a quantity can be processed. Throughput assesses the effectiveness of an entire operation or process by measuring speed of production. Throughput is based on the ratio of the number of outputs to the operating time (Figure 7.35).

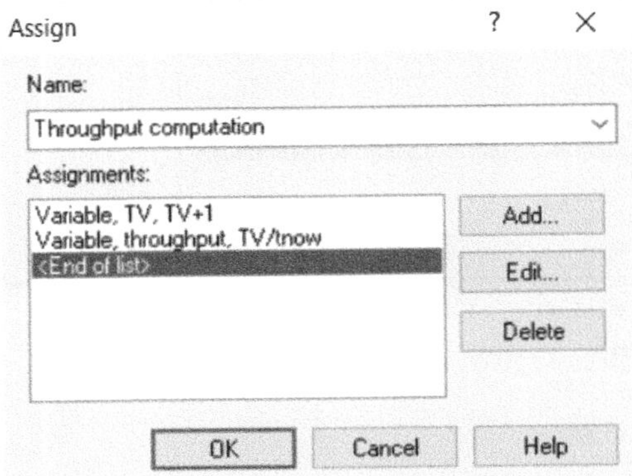

Figure 7.35. *Configuration of the “Assign” module “Throughput computation”, television assembly line*

7.1.4.4. *Modeling of operating schedules*

The “Resource” module ensures the modeling of assembly line operating schedules. The type of resource must be “Based on Schedule”, which means aligned with a calendar (Figure 7.36). In the context of this project, the calendar is called “Operating schedule”.

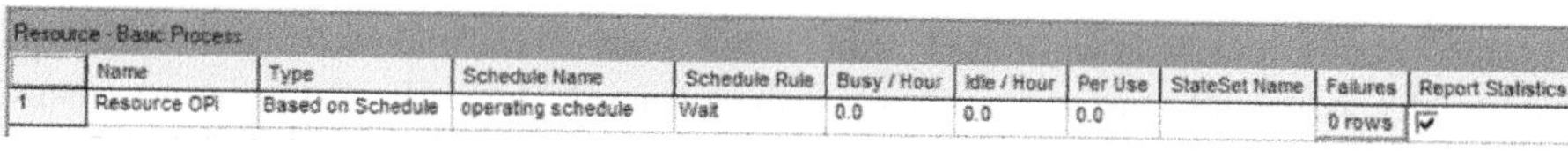

Resource - Basic Process

	Name	Type	Schedule Name	Schedule Rule	Busy / Hour	Idle / Hour	Per Use	StateSet Name	Failures	Report Statistics
1	Resource OPi	Based on Schedule	operating schedule	Wait	0.0	0.0	0.0		0 rows	☑

Figure 7.36. *Configuration of the "Resource" module assigned to the "Process" module, television assembly line*

The implementation of the "Operating schedule" is effected through the "Schedule" module. This calendar operates under the "Capacity" type, using a time unit equivalent to a quarter of an hour. The task then involves assigning the resource capacity according to the time units using the "Durations" table. Here, the "Value" column corresponds to the number of resources, while the "Durations" column corresponds to the respective number of quarter hours. For instance, during the first quarter hour, no resources are allocated due to the operator delay from 8 a.m. to 8:15 a.m. Subsequently, an operator is assigned to the "Process" module during the 7 quarter hours and so on (Figure 7.37). The data in this table are derived from the schedule given in Table 7.5.

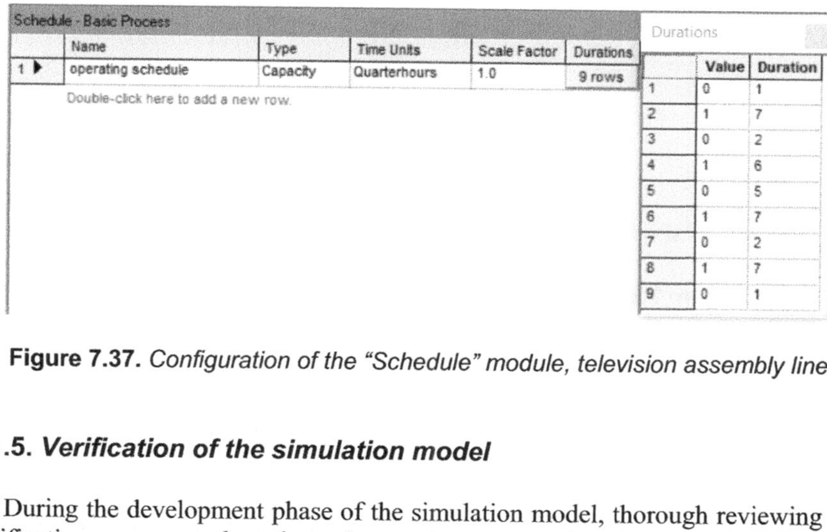

Schedule - Basic Process

	Name	Type	Time Units	Scale Factor	Durations
1	operating schedule	Capacity	Quarterhours	1.0	9 rows

Double-click here to add a new row.

Durations

	Value	Duration
1	0	1
2	1	7
3	0	2
4	1	6
5	0	5
6	1	7
7	0	2
8	1	7
9	0	1

Figure 7.37. *Configuration of the "Schedule" module, television assembly line*

7.1.5. *Verification of the simulation model*

During the development phase of the simulation model, thorough reviewing and verification were conducted to insure its proper functioning. An additional verification method was used, involving the adjustment of stochastic values of assembly operation times to a constant value of 5 seconds. The first release of a television from the assembling line could therefore be anticipated to occur at:

$$t=\underbrace{28*5}_{\text{Assembly operations}}+\underbrace{28*4}_{\text{Transfer}}+\underbrace{15*60}_{\text{First 15 minutes lost at start of day}}=1{,}152 \text{ seconds.}$$

This first output was validated after simulation. The final verification method employed was model animation. This involves tracing the path of entities to ensure alignment with the conceptual model. These three methods collectively confirmed the verification of the simulation model.

7.1.6. *Validation of the simulation model*

7.1.6.1. *Transition period*

A sample simulation of 12 days of operation was conducted with throughput rate plotted accordingly. As depicted in Figure 7.38, the rate began to stabilize after two days of operation. Hence, this initial two-day phase was considered as the transitional period and is subsequently excluded from the simulation results.

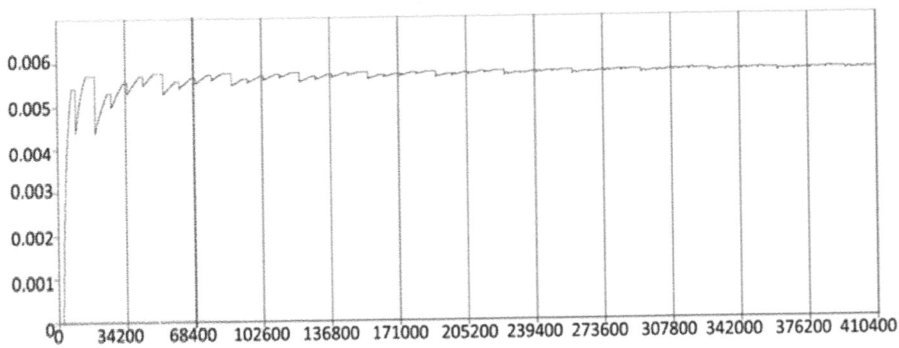

Figure 7.38. *Throughput rate of the television assembly line*

7.1.6.2. *Number of replications*

The determination of the required number of replications relies on the analysis of the outcomes of the initial 10 replications. Each of these replications spans 12 days of operation, with a two day transitional period. Following the completion of each replication, the statistics are reset, and the simulation model starts empty. The statistic gathered during these simulation replications was the number of televisions assembled (Table 7.6).

To determine the necessary number of replications, the mean and standard deviation for these first 10 replications should be computed. The average of the 10 values is 1,973.9 TVs, and the standard deviation of the sample is 1.663 TVs. Calculating the confidence interval yields CI = 1.188 TVs, given that $t_{1\text{-}0.05/2,\,10\text{-}1} = 2.26$. This confidence interval is sufficiently precise, and the analysis of the simulation results does not require additional replications.

Replication	Mounted TVs	Replication	Mounted TVs
1	1,976	6	1,974
2	1,972	7	1,977
3	1,973	8	1,974
4	1,973	9	1,972
5	1,973	10	1,975

Table 7.6. *Result of 10 simulation replications, television assembly line*

7.1.6.3. *Validation*

To validate the simulation model of the assembly line, it is imperative to compare the actual results of the line with those generated by the simulated model. Table 7.7 presents the results of 10 simulation replications, as well as the statistics for the real assembly line.

Replication	Simulation	Real	Replication	Simulation	Real
1	1,976	1,980	6	1,974	1,977
2	1,972	1,965	7	1,977	1,981
3	1,973	1,982	8	1,974	1,979
4	1,973	1,974	9	1,972	1,972
5	1,973	1,975	10	1,975	1,970

Table 7.7. *Results of simulation model and real assembly line*

The curves corresponding to these values were meticulously plotted (Figure 7.39).

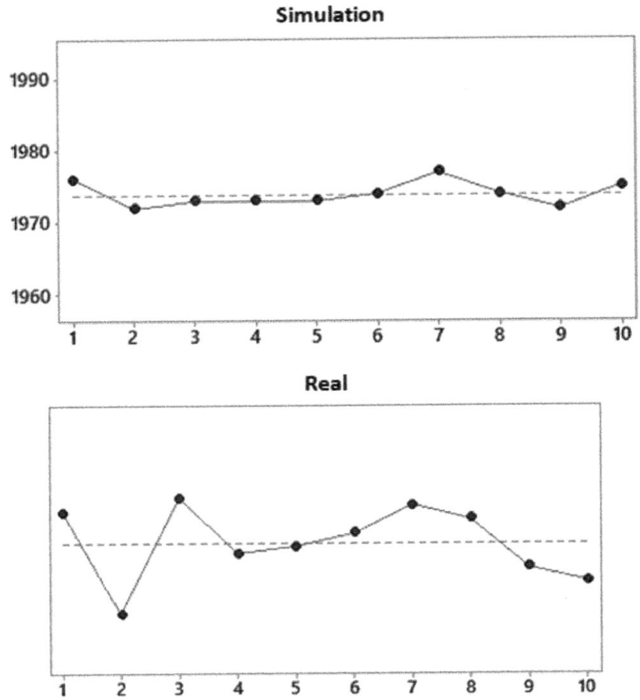

Figure 7.39. *Curves for the number of televisions assembled by the real assembly line and in the corresponding simulation model*

Analyzing the results of these 10 iterations enabled the modeling of these values within confidence intervals, as depicted in Table 7.8.

Statistical	Simulation	Real
Average	1,973.9	1,975.5
Standard deviation	1.663	5.40
Confidence interval	[1,973; 1,975]	[1,971.6; 1,979.4]

Table 7.8. *Means, standard deviations and confidence intervals for the number of TVs assembled*

These confidence intervals, illustrated in Figure 7.40, reveal overlapping ranges, implying equivalence in the results.

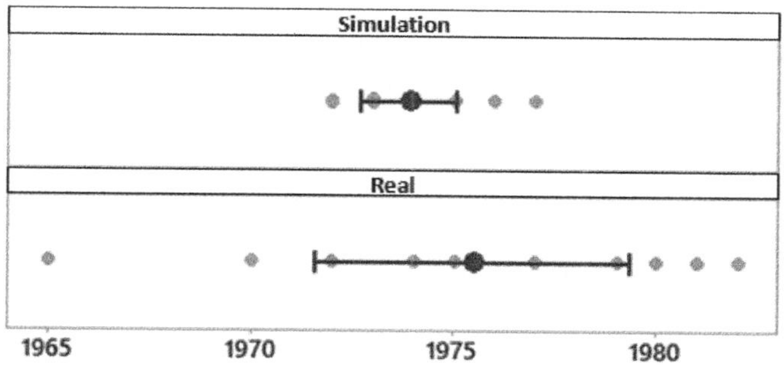

Figure 7.40. *Confidence intervals for the number of TVs assembled by the real assembly line and in the corresponding simulation model. For a color version of the figure, see www.iste.co.uk/jerbi/discrete.zip*

Application of Student's *t*-test was necessary to confirm this equivalence. This test evaluates whether the difference between the total number N of televisions assembled by the real chain and that in the simulation model is statistically significant or not. Employing the *t*-test at a significance level of 5%, the following hypotheses were considered:

– H_0: difference = $N_{real} - N_{model} = 0$;

– H_1: difference = $N_{real} - N_{model} \neq 0$.

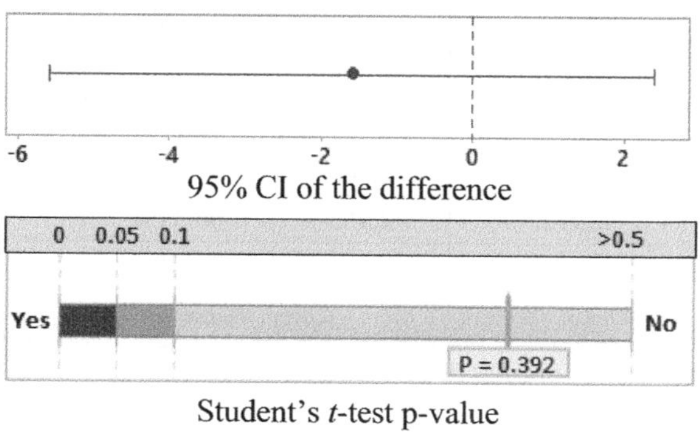

Figure 7.41. *Validation of the simulation model. For a color version of the figure, see www.iste.co.uk/jerbi/discrete.zip*

The confidence interval of the difference between N_{real} and N_{model} quantifies the uncertainty associated with estimating the difference of means from sample data. As illustrated in Figure 7.41, it is 95% confident that the real difference is between

–5.58 and 2.38. This confidence interval therefore includes the value of 0. In addition, the p-value of Student's test is 0.392, which exceeds 0.05. Consequently, these results indicate that there is not sufficient evidence to conclude that the means differ at the 0.05 level of significance. Hence, the simulation model of the television assembly line is valid.

7.1.7. *Experimentation and analysis of results*

7.1.7.1. *Current state*

7.1.7.1.1. Assembly line efficiency

The efficiency of the TV assembly line can be assessed by examining the ratio of televisions assembled to those arriving on the line. A higher ratio signifies a more efficient chain. As indicated in Table 7.9, the chain efficiency is relatively satisfactory. Nearly 66% of entries are assembled in 12 days. The result is confirmed by the relatively low number for work in progress on the assembly line.

Replication	Assembled TVs	Arrivals	Ratio	Work in progress
1	1,976	2,968	66.58%	710.07
2	1,972	2,964	66.53%	705.69
3	1,973	2,968	66.48%	709.06
4	1,973	2,967	66.50%	706.91
5	1,973	2,966	66.52%	706.63
6	1,974	2,966	66.55%	709.37
7	1,977	2,964	66.70%	706.25
8	1,974	2,966	66.55%	706.29
9	1,972	2,965	66.51%	707.30
10	1,975	2,965	66.61%	705.39

Table 7.9. *Ratio of the number of televisions assembled in relation to the number of entries to the assembly line and work in progress*

7.1.7.1.2. Search for bottleneck positions

This investigation is conducted by examining the average number of products waiting in each workstation (Table 7.10).

The simulation results reveal that most workstations do not have a significant number of waiting products. However, workstations OP14, OP15 and OP23 stand out with notably high averages for waiting products.

Rep	1	2	3	4	5	6	7	8	9	10	Avg
OP1	8.01	7.98	8.02	7.99	7.99	7.99	8.03	8.00	8.00	7.97	8.00
OP2	0.69	0.91	0.49	0.55	1.29	0.92	0.92	1.12	1.17	0.99	0.91
OP3	0.47	0.54	0.46	0.49	0.51	0.47	0.45	0.50	0.49	0.49	0.49
OP4	0.48	0.10	0.31	0.65	0.30	0.06	0.35	0.23	0.32	0.29	0.31
OP5	0.02	0.23	0.05	0.00	0.09	0.17	0.02	0.16	0.00	0.01	0.07
OP6	0.00	0.00	0.00	0.01	0.00	0.01	0.01	0.00	0.16	0.00	0.02
OP7	1.81	1.70	1.77	1.75	1.81	1.70	1.84	1.76	1.78	1.73	1.76
OP8	1.01	1.06	1.06	1.08	1.12	1.10	0.99	1.02	1.14	0.99	1.06
OP9	0.88	0.87	0.91	0.88	0.79	0.75	0.81	0.80	0.78	0.87	0.83
OP10	0.02	0.06	0.02	0.02	0.01	0.02	0.03	0.05	0.02	0.02	0.03
OP11	0.02	0.03	0.03	0.02	0.02	0.02	0.02	0.02	0.02	0.03	0.02
OP12	5.77	5.72	7.00	5.94	5.37	5.19	5.53	5.25	5.55	6.25	5.76
OP13	0.92	1.31	1.03	1.68	1.49	1.58	1.17	1.27	1.40	1.12	1.30
OP14	**96.96**	**92.70**	**92.22**	**94.95**	**96.30**	**95.01**	**91.63**	**97.63**	**91.68**	**92.20**	**94.13**
OP15	**153.60**	**162.28**	**157.36**	**158.85**	**151.43**	**158.07**	**157.58**	**149.57**	**153.56**	**157.14**	**155.95**
OP16	3.28	2.85	3.09	3.16	3.33	3.01	2.89	3.36	3.04	2.75	3.08
OP17	0.57	0.44	0.50	0.57	0.60	0.46	0.42	0.65	0.47	0.40	0.51
OP18	0.02	0.00	0.00	0.00	0.02	0.03	0.00	0.00	0.03	0.06	0.02
OP19	0.11	0.00	0.07	0.01	0.00	0.01	0.24	0.03	0.00	0.10	0.06
OP20	0.02	0.06	0.10	0.11	0.06	0.01	0.01	0.01	0.09	0.00	0.05
OP21	0.01	0.01	0.01	0.01	0.02	0.01	0.01	0.04	0.06	0.01	0.02
OP22	0.57	0.39	0.36	0.55	0.30	0.47	0.07	0.46	0.31	0.45	0.39
OP23	**419.03**	**410.75**	**418.31**	**411.57**	**417.76**	**416.29**	**417.37**	**418.72**	**421.86**	**415.49**	**416.71**
OP24	2.65	2.62	3.02	2.80	2.66	3.09	2.86	2.72	2.35	3.01	2.78
OP25	0.02	0.22	0.04	0.24	0.03	0.04	0.03	0.03	0.11	0.26	0.10
OP26	0.47	0.34	0.27	0.40	0.70	0.34	0.40	0.35	0.39	0.19	0.39
OP27	0.09	0.04	0.03	0.11	0.07	0.01	0.02	0.00	0.00	0.01	0.04
OP28	0.02	0.02	0.02	0.02	0.02	0.02	0.02	0.02	0.02	0.06	0.02

Table 7.10. *Number of products on hold per workstation*

7.1.7.2. *Investigation of the effect of staff delay*

To investigate the effect of staff delays on line production, the attendance schedule for all staff is adjusted to respect the normal operating time. Table 7.11 shows the number of TVs assembled under both operating schedules, with and without delay.

Replication	With delay	Without delay	Replication	With delay	Without delay
1	1,976	2,339	6	1,974	2,339
2	1,972	2,340	7	1,977	2,336
3	1,973	2,341	8	1,974	2,339
4	1,973	2,336	9	1,972	2,338
5	1,973	2,337	10	1,975	2,335

Table 7.11. *Number of televisions assembled, schedules with and without delay*

To compare these results, a hypothesis test can be used. This test determines the difference between the number of televisions assembled by the assembly line with and without staff delays. This test is conducted at a significance level of 5% with the following hypothesis:

– H_0: difference = $N_{with\_delay} - N_{without\_delay} = 0$;

– H_1: difference = $N_{with\_delay} - N_{without\_delay} \neq 0$.

The confidence interval of the difference quantifies the uncertainty associated with estimating the difference of means from sample data. As shown in Table 7.12, there is 95% certainty that the difference is between –365.81 and –362.39 TVs. This confidence interval does not include the value of 0. In addition, the p-value of the Student's test is less than 0.001, indicating that the means are significantly different at a 0.05 significance level. This results in an 18.44% improvement in the number of televisions assembled by the assembly line when the schedule has no delays. Conversely, simulation results indicate that only assembly workstation OP23 has very high averages for waiting products (Table 7.13).

Statistical	With delay	Without delay
Average	1,973.9	2,338
Standard deviation	1.66	1.94
95% CI	[1,973; 1,975]	[2,336.6; 2,339.4]
95% CI difference	[–365.81; –362.39]	
P value	P<0.001	

Table 7.12. *Comparison between numbers of televisions assembled for schedules with and without delay*

Rep	1	2	3	4	5	6	7	8	9	10	Avg
OP1	3.92	3.88	3.89	3.89	3.90	3.92	3.91	3.89	3.87	3.88	3.90
OP2	0.42	0.47	0.53	0.43	0.59	0.42	0.44	0.53	0.62	0.55	0.50
OP3	0.22	0.25	0.25	0.25	0.24	0.22	0.25	0.26	0.23	0.26	0.24
OP4	0.21	0.15	0.24	0.19	0.16	0.18	0.16	0.16	0.16	0.29	0.19
OP5	0.15	0.00	0.00	0.00	0.11	0.11	0.16	0.05	0.23	0.04	0.09
OP6	0.00	0.15	0.00	0.00	0.00	0.00	0.01	0.00	0.00	0.00	0.02
OP7	0.56	0.56	0.58	0.54	0.58	0.55	0.57	0.57	0.57	0.56	0.56
OP8	0.30	0.29	0.29	0.29	0.27	0.29	0.28	0.28	0.28	0.30	0.29
OP9	0.30	0.32	0.33	0.39	0.37	0.34	0.37	0.34	0.32	0.32	0.34
OP10	0.01	0.01	0.01	0.01	0.01	0.01	0.02	0.01	0.01	0.01	0.01
OP11	0.01	0.01	0.01	0.01	0.01	0.01	0.01	0.01	0.01	0.01	0.01
OP12	1.37	1.37	1.42	1.34	1.47	1.45	1.43	1.43	1.34	1.44	1.41
OP13	0.25	0.36	0.30	0.40	0.27	0.53	0.30	0.19	0.43	0.25	0.33
OP14	1.92	1.87	1.92	2.00	1.84	2.03	1.88	1.82	2.01	1.77	1.91
OP15	4.42	4.40	4.68	4.25	4.52	4.64	4.26	4.28	4.48	4.25	4.42
OP16	0.69	0.65	0.63	0.69	0.59	0.59	0.61	0.69	0.64	0.64	0.64
OP17	0.06	0.05	0.06	0.06	0.05	0.05	0.05	0.07	0.05	0.05	0.06
OP18	0.00	0.00	0.00	0.00	0.03	0.03	0.02	0.00	0.00	0.00	0.01
OP19	0.01	0.09	0.12	0.02	0.00	0.06	0.04	0.02	0.09	0.00	0.05
OP20	0.00	0.00	0.00	0.00	0.10	0.12	0.01	0.03	0.00	0.00	0.03
OP21	0.00	0.02	0.03	0.03	0.00	0.00	0.01	0.00	0.00	0.02	0.01
OP22	0.28	0.09	0.17	0.22	0.18	0.10	0.31	0.26	0.32	0.41	0.24
OP23	**421.79**	**420.95**	**419.11**	**424.29**	**420.95**	**422.99**	**423.57**	**425.34**	**425.04**	**425.40**	**422.94**
OP24	1.27	1.22	1.15	1.24	1.16	1.11	1.04	0.99	1.24	1.10	1.15
OP25	0.02	0.01	0.09	0.01	0.01	0.17	0.16	0.15	0.11	0.01	0.07
OP26	0.21	0.18	0.28	0.17	0.23	0.15	0.07	0.24	0.11	0.18	0.18
OP27	0.00	0.04	0.00	0.00	0.03	0.01	0.00	0.03	0.00	0.03	0.01
OP28	0.01	0.01	0.01	0.01	0.01	0.01	0.01	0.01	0.01	0.01	0.01

Table 7.13. *Number of products waiting per workstation, schedule without delay*

Staff failure to adhere to operating hours significantly impacts the performance of the two other workstations OP14 and OP15. Respecting the operating time alone was sufficient to reduce the number of products waiting in these workstations.

7.1.7.3. *Increase in the number of staff in the congested workstation*

To enhance the performance of the television assembly line, a study was conducted to examine the effect of increasing the number of OP23 workstations. The approach involves adding a second OP23 workstation, bringing the total to two. This is implemented using two "Process" modules named "OP23a" and "OP23b", coupled with two "Station" modules and two "Route" modules. These modules are configured to direct the entities to the "OP24" station after processing at either of the "OP23i" workstations. The transfer from the "OP22" workstation to one of the "OP23i" workstations is managed by a "PickStation" module, which selects the available "OP23i" workstation with the lowest waiting entities (Figure 7.42).

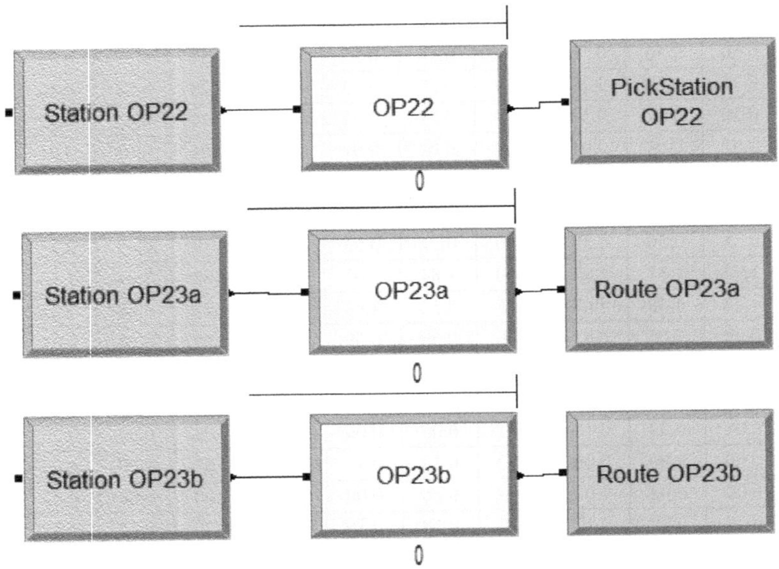

Figure 7.42. *Increase in the number of OP23 workstations. For a color version of the figure, see www.iste.co.uk/jerbi/discrete.zip*

The simulation results show that the number of televisions assembled by the assembly line increases significantly (Table 7.14).

Replication	Without delay	Without delay and two OP23s	Replication	Without delay	Without delay and two OP23s
1	2,339	2,965	6	2,339	2,958
2	2,340	2,960	7	2,336	2,963
3	2,341	2,961	8	2,339	2,965
4	2,336	2,963	9	2,338	2,963
5	2,337	2,963	10	2,335	2,962

Table 7.14. *Number of televisions mounted, schedule without delay with two operators at workstation OP23*

From the perspective of the number of waiting entities, the simulation results indicate that adding a second OP23 workstation completely decongests the assembly line (Table 7.15).

Rep	1	2	3	4	5	6	7	8	9	10	Avg
OP1	3.90	3.88	3.90	3.88	3.87	3.90	3.90	3.87	3.88	3.89	3.89
OP2	0.45	0.77	0.62	0.60	0.46	0.46	0.62	0.53	0.56	0.36	0.54
OP3	0.25	0.27	0.26	0.24	0.24	0.26	0.24	0.26	0.23	0.23	0.25
OP4	0.07	0.15	0.22	0.07	0.18	0.18	0.16	0.15	0.25	0.21	0.16
OP5	0.13	0.01	0.04	0.11	0.02	0.03	0.10	0.00	0.14	0.00	0.06
OP6	0.03	0.10	0.11	0.00	0.00	0.13	0.00	0.02	0.01	0.15	0.05
OP7	0.55	0.55	0.58	0.56	0.56	0.53	0.56	0.57	0.55	0.56	0.56
OP8	0.29	0.28	0.26	0.25	0.28	0.28	0.26	0.27	0.31	0.26	0.28
OP9	0.35	0.31	0.30	0.35	0.38	0.27	0.41	0.36	0.34	0.27	0.33
OP10	0.01	0.00	0.00	0.01	0.01	0.00	0.01	0.01	0.01	0.02	0.01
OP11	0.01	0.01	0.01	0.01	0.01	0.01	0.01	0.01	0.01	0.01	0.01
OP12	1.71	1.45	1.43	1.41	1.39	1.51	1.43	1.38	1.36	1.34	1.44
OP13	0.32	0.28	0.30	0.28	0.42	0.33	0.25	0.34	0.29	0.33	0.31
OP14	2.02	1.87	1.91	1.74	1.93	1.97	1.86	1.85	1.78	1.81	1.88
OP15	4.58	4.88	4.57	4.29	4.55	4.64	4.29	4.55	4.35	4.37	4.51
OP16	0.62	0.65	0.57	0.67	0.65	0.58	0.52	0.67	0.63	0.67	0.62
OP17	0.07	0.07	0.05	0.05	0.05	0.05	0.03	0.06	0.06	0.06	0.06
OP18	0.00	0.03	0.01	0.00	0.05	0.01	0.00	0.00	0.00	0.00	0.01
OP19	0.13	0.10	0.03	0.00	0.01	0.01	0.18	0.03	0.09	0.04	0.06
OP20	0.00	0.00	0.01	0.01	0.01	0.00	0.05	0.00	0.02	0.00	0.01
OP21	0.00	0.06	0.00	0.00	0.00	0.15	0.00	0.00	0.01	0.00	0.02
OP22	0.16	0.23	0.45	0.26	0.21	0.06	0.26	0.35	0.24	0.11	0.23
OP23a	0.01	0.01	0.02	0.01	0.01	0.01	0.01	0.02	0.01	0.01	0.01
OP23b	0.00	0.00	0.00	0.00	0.00	0.01	0.00	0.00	0.00	0.00	0.00
OP24	0.46	0.34	0.06	0.27	0.36	0.35	0.22	0.18	0.19	0.35	0.28
OP25	0.00	0.00	0.14	0.02	0.00	0.00	0.01	0.00	0.04	0.00	0.02
OP26	0.21	0.23	0.20	0.23	0.24	0.33	0.26	0.33	0.29	0.29	0.26
OP27	0.00	0.01	0.09	0.00	0.00	0.02	0.00	0.04	0.00	0.07	0.02
OP28	0.01	0.01	0.00	0.02	0.01	0.01	0.02	0.01	0.01	0.01	0.01

Table 7.15. *Number of products waiting per workstation, schedule without delay with two OP23 workstations*

7.1.8. *Conclusion*

The study of this television assembly line highlights the significant impact of staff adhering to operating times. Indeed, this helps to decongest the assembly line by reducing the number of products waiting in certain workstations. However, to fully address congestion, some positions need to be doubled. These similar workstations operate in parallel and contribute to increasing the number of televisions assembled by the assembly line.

7.2. Simulation-based optimization of a flexible manufacturing system

The uncertain nature of current markets has spurred the development of manufacturing systems, leading to the creation of systems with varied principles and functionalities designed to meet constantly changing demands. Manufacturing systems now operate in environments where product life cycles are increasingly short, demands are highly unpredictable and products feature customizable options. This challenging environment has compelled companies to aim for high levels of responsiveness and flexibility to ensure they meet customer needs effectively.

Generally, optimizing the performance of a flexible manufacturing system (FMS) can be achieved through various methods. The most commonly used are exact methods and approximate methods. These methods often face challenges due to the necessity of developing complex mathematical models that accurately reflect the impact of one or more factors on the performance of the FMS.

7.2.1. *Objective*

The objective of this project is to optimize the mean flow time (MFT) of parts in an FMS using simulation by varying several key factors. MFT represents the average time it takes for a batch of parts to be manufactured in the system, serving as an indicator of the fluidity and efficiency of a part's flow through the FMS.

7.2.2. *The FMS studied*

7.2.2.1. *The actual model*

The system comprises eight machines that process four types of products grouped into two families based on their similarity. Each product requires between two and five production operations. The system is organized in a functional layout, also known as a "Job shop" layout, where machines are arranged into departments based on their type. In this study, the machines are grouped into three departments (Figure 7.43).

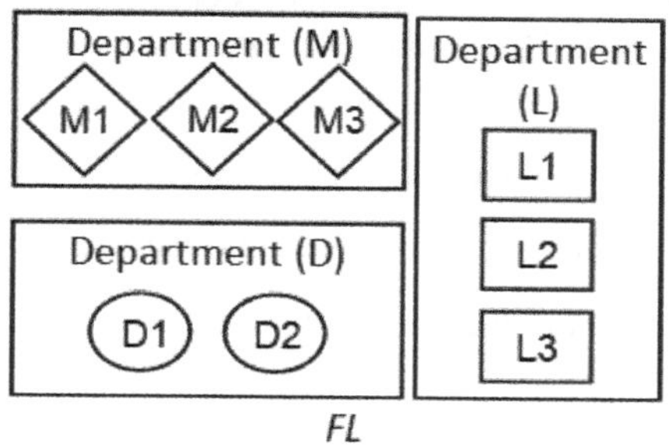

Part	Routing	Family
P1	L→M→D	2
P2	L→D→M	1
P3	L→M	1
P4	L→M→D→L→M	2

Figure 7.43. *Structure of the FMS studied*

Products arrive at the FMS with an inter-arrival time (IAT) of 15 minutes per unit and are grouped into batches of five units for transfer between departments. The transfer time (TT) follows a uniform distribution with U (5, 8) minutes. Before processing by a department's machine, batches wait in a queue controlled by the first come first served (FCFS) priority rule, which processes batches according to their order of arrival. The manufacturing time (PT) and machine setup time (ST) required for each production operation are detailed in Table 7.16.

		L	M	D	L	M
P1	ST	T(39, 44, 49)	U(49, 69)	N(65, 15)	–	–
	P.T.	T(15, 18, 21)	U(9, 13)	N (29, 5)	–	–
P2	ST	T (90, 101, 110)	U(64, 84)	N(107 35)	–	–
	P.T.	T (17, 21, 25)	U(20, 28)	N(14, 5)	–	–
P3	ST	T (80, 84, 88)	U(72, 92)	–	–	–
	P.T.	T(23, 28, 32)	U (14, 18)	–	–	–
P4	ST	T (62, 66, 70)	U(50, 58)	N(101, 10)	T(33, 38, 45)	U(78, 98)
	P.T.	T(18, 20, 22)	U(25, 33)	N(12, 5)	T(20, 23, 26)	U(15, 23)

Notes: N(A, SD): normal distribution with mean A and standard deviation SD. U (Min, Max): uniform distribution between Min and Max. T(Min, Mode, Max): triangular distribution between Min, Mode and Max. All times are in minutes.

Table 7.16. *Process and setup times for production operations in the FMS*

Based on the similarity between the products entering and leaving the machines, the setup time can be reduced by applying a reduction factor λ. This factor can take on one of the following three values:

– $\lambda = 0$ if the successive products are of the same type;

– $\lambda = 0.8$ if the two products are from the same family, but of two different types;

– $\lambda = 1$ if the successive products are from different families.

7.2.2.2. *The models explored*

One potential optimization of the system under study is to adopt a cellular layout. This layout is based on the principle of group technology, which focuses on similar and repetitive activities. In a cellular layout, similar products are grouped into families, and the machines required for all the production phases of each family are assembled into cells. Each cell must be autonomous; otherwise, inter-cell transfers can degrade the performance of the cell layout. Converting the current functional layout of the FMS into a cellular layout involves grouping the machines into two cells, each processing one of the two product families (Figure 7.44).

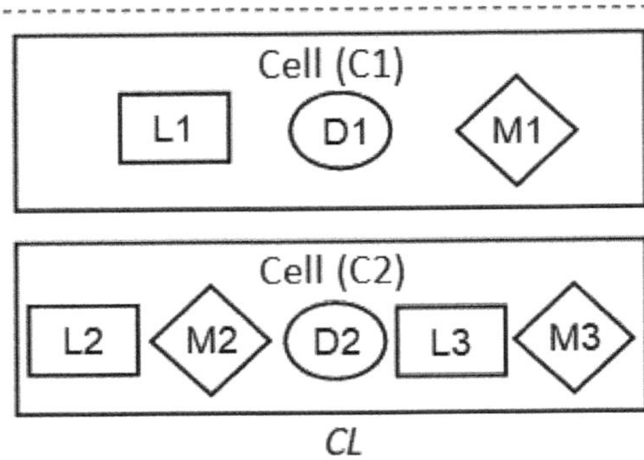

Part	Routing	Family
P1	L2→M2→D2	2
P2	L1→D1→M1	1
P3	L1→M1	1
P4	L2→M2→D2→L3→M3	2

Figure 7.44. *Structure of the FMS in cellular layout*

Another mode of product transfer under consideration involves transferring products individually rather than in batches. In this transfer mode, each product moves to the next machine or department without waiting for the rest of its batch to be processed. Implementing this transfer mode require a different queue priority rule known as shortest processing time (SPT). This rule prioritizes products or batches with the lowest production time.

All factors examined and their corresponding levels used in this study are summarized in Table 7.17.

Factor	Levels	
	1	**2**
Layout	FL	CL
Inter-arrival time (IAT minutes/piece)	15	25
Batch size (BS)	5	10
Process time (PT in minutes)	80% of level 2	Values from Table 7.16
Setup time (ST in minutes)	80% of level 2	Values from Table 7.16
Transfer time (TT in minutes)	U (5, 8)	U (10, 16)
ST reduction factor (λ)	0.5	0.8
Queue rule (rule)	FCFS	SPT
Overlap transfer	With	Without

Note: U (Min, Max): uniform distribution between Min and Max.

Table 7.17. *Levels of FMS factors*

7.2.3. *Simulation models*

7.2.3.1. *Functional layout simulation model*

7.2.3.1.1. Arrival of products

The arrival of products P_1, P_2, P_3 and P_4 is managed through a "Create" module named "Product Creation" where the arrival times and batch sizes are specified. This module is followed by a "Batch" module that groups the products into batches. These batches then pass through an "Assign" module named "Performance index entry", where the "types" attribute is assigned to each product according to a discrete distribution. Each product has a 25% chance of belonging to one of the four product types. Next, the "Separate" module, named "Product separation", separates the batches back into individual products. A "Decide" module, named "Product entry sorting", then sorts the products by type before they pass through one of the four "Assign" modules, named "Product i assignments". These modules assign necessary attributes such as PT, ST, product family and the names of each production sequence. The specific steps for each sequence are defined in "Sequence" modules. Finally, the products are transferred to one of the departments via a "Route" module named "Transfer to departments" (Figure 7.45).

Two simulation models of the functional layout were developed, each corresponding to a different mode of product transfer. The first is illustrated in Figure 7.45, and based on the "overlapping" transfer mode, it transfers products individually. The second model, depicted in Figure 7.46 and based on the "nooverlapping" transfer mode, transfers products in batches. This model includes an additional "Batch" module, named "Product regrouping", which groups the products into batches before they pass through the "Route" module.

7.2.3.1.2. The departments

This section consists of three departments, each comprising two components: the queue and the machines. When a product or a batch of products arrives at one of the departments, it is placed in the department's queue via the "Hold" module named "Department i queue" until one of the machines becomes available. The selection of products from the queue follows a predetermined priority rule, which is defined through the "Queue" modules. Once a product is selected, it is released by the "Hold" module and undergoes a decision process via the "Decide" module, named "Choice of available machine dep i", which assigns it to the available machine. Figure 7.47 presents the model of a department using the "overlapping" transfer mode.

In the case of the "nooverlapping" transfer mode using the FCFS control rule, the batches of products are separated via the "Separate" module named "Separation products dep i". After processing by the department's machines, the products are regrouped by the "Batch" module named "Product regrouping dep i" following the machines (Figure 7.48). In the functional model with the "nooverlapping" transfer mode using the SPT control rule, incoming product batches are held in an initial "Hold" module until selected. Once selected, the batch is separated into individual products by a "Separate" module named "Separation products dep i", and then put on hold in a second "Hold" module named "Buffer hold dep i". This second hold module acts as a temporary queue from which the products are transferred to the department's available machines. This process ensures the entire selected batch of products is processed (Figure 7.49). Similar to the previous case, the processed products are grouped into batches when they exit the department's machines and are then transferred in batches to the next department via the "Route" module.

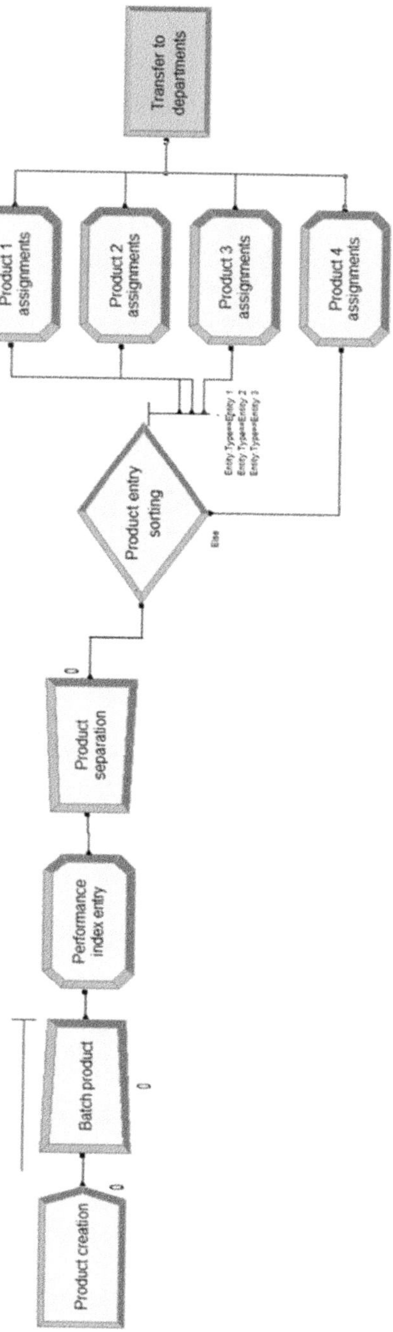

Figure 7.45. *Arrival of products with the "overlapping" transfer mode. For a color version of the figure, see www.iste.co.uk/jerbi/discrete.zip*

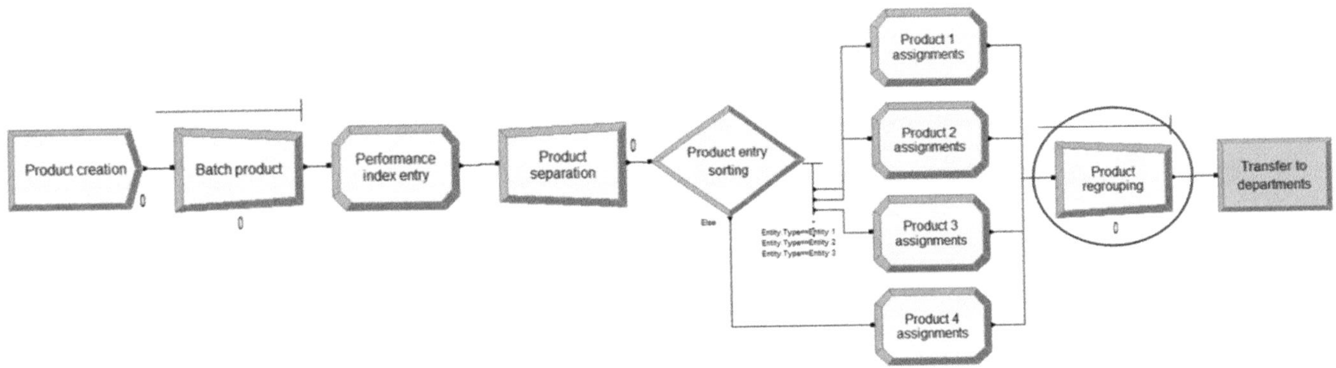

Figure 7.46. *Arrival of products with the non-overlapping transfer mode "nooverlapping"*

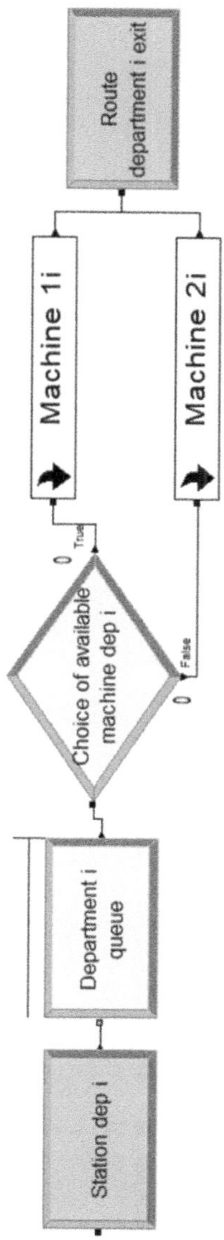

Figure 7.47. *Structure of a department with the overlapping transfer mode. For a color version of the figure, see www.iste.co.uk/jerbi/discrete.zip*

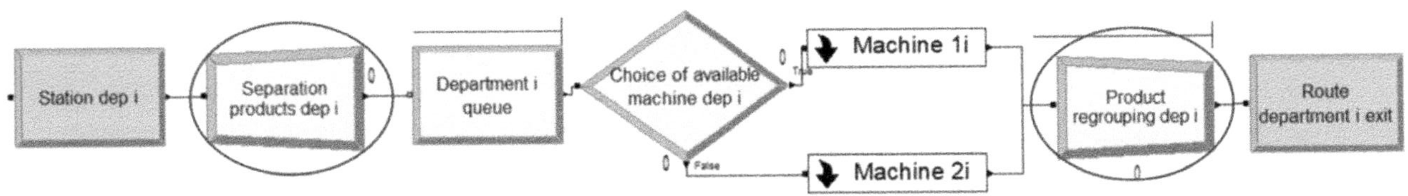

Figure 7.48. *Functional model department structure with the "nooverlapping" transfer mode and FCFS control rule. For a color version of the figure, see www.iste.co.uk/jerbi/discrete.zip*

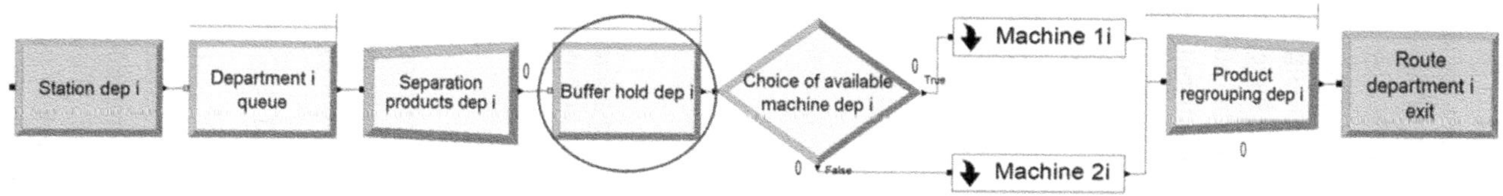

Figure 7.49. *Functional model department structure with the "nooverlapping" transfer mode and SPT control rule. For a color version of the figure, see www.iste.co.uk/jerbi/discrete.zip*

Unlike all other products, the P_4 product has a particular processing requirement. This product necessitates two different production operations, which must be performed by departments L and M. Consequently, it involves two different processing and setup times, which must be used by these departments. To address this constraint, only the processing and setup times of the first three operations are allocated to product P_4 in the "Product 4 assignment" module located at the product entry point. Once the P_4 type products have been processed by the machines of department Di, they are sorted among the other types of products via a "Decide" module called "Sorting P4 for times reattribution". These products are then directed to a new "Assign" module, named "Times reattribution for P4", where the new processing and setup times required for the two subsequent operations in departments L and M are allocated (Figure 7.50).

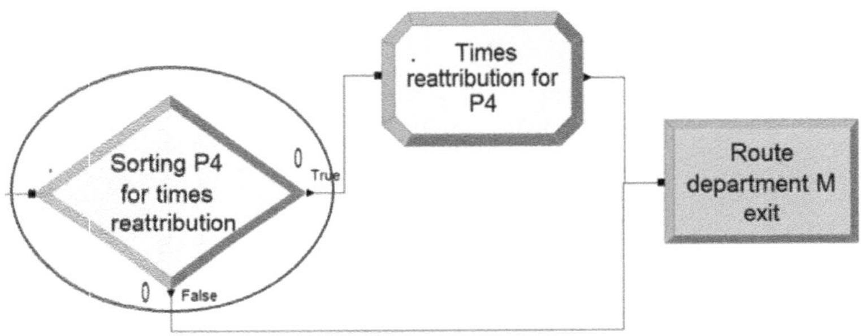

Figure 7.50. *New production and adjustment times for the P4 product. For a color version of the figure, see www.iste.co.uk/jerbi/discrete.zip*

7.2.3.1.3. Machines

The machines in each department are modeled using "Process" modules. Within these modules, the transformation times are specified as a function of the production time (PT) and the setup time (ST) weighted by the lambda factor (λ):

$$\text{Transformation time} = PT + \lambda * ST \qquad [7.1]$$

The value of the factor λ depends on the similarity of the types of products entering and leaving the machine. A "Decide" module, named "Decide Mij", tests all products to determine this factor. It compares two variables defined via the "Assign" modules "type in Mij" and "type out Mij". If these variables are identical, the "Decide Mij" module directs the incoming product to the corresponding "Assign" module, "Lamda 0 Mij", setting λ to 0. When the products are of the same family but different types, the "Decide Mij" module directs the product "Lamda 0.5 Mij" module, setting λ to 0.5. If the products belong to different families, the "Decide Mij" module directs the product to the "Lamda 1 Mij" module, setting λ to 1 (Figure 7.51).

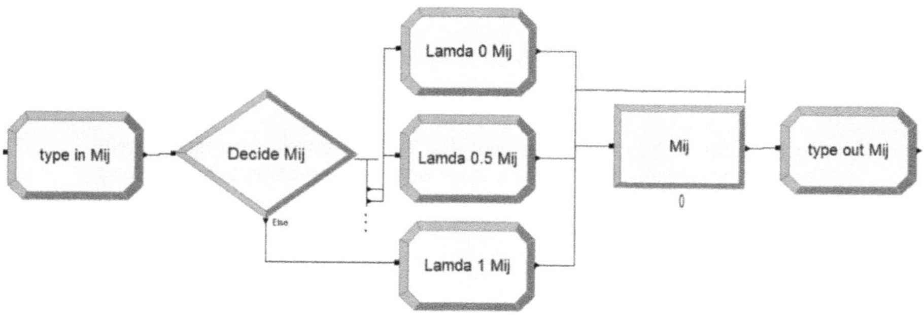

Figure 7.51. *Modeling of FL layout machines. For a color version of the figure, see www.iste.co.uk/jerbi/discrete.zip*

7.2.3.1.4. Product release

All products that have completed manufacturing are transferred to the product exit section. This third section of the simulation model begins with a "Station" connected to a "Batch" module that groups the products into batches according to their types, especially in the "overlapping" transfer mode, before being evacuated from the simulation model via the "Dispose" module named "System exit" (Figure 7.52).

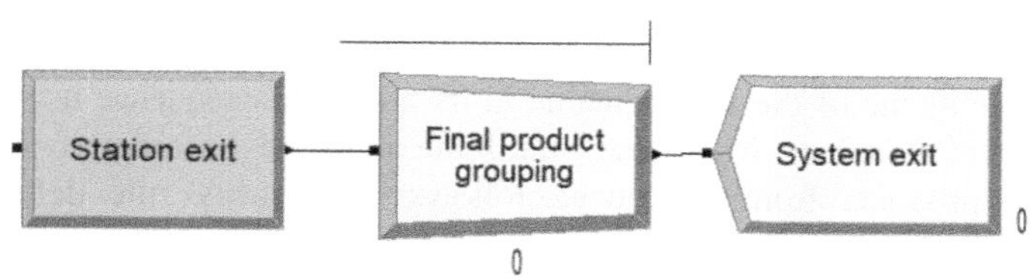

Figure 7.52. *Product release with transfer mode over. For a color version of the figure, see www.iste.co.uk/jerbi/discrete.zip*

In the case of the "nooverlapping" transfer mode, batches of products are separated by a "Separate" module named "Final separation of products" and then regrouped by a "Batch" module called "Final groupings of products" (Figure 7.53). This is necessary because the "Dispose" module only accepts batches of the "Permanent" type. When entering the exit section the batches are of the "Temporary" type due to the multiple dissociations that occur during processing by the different functional departments.

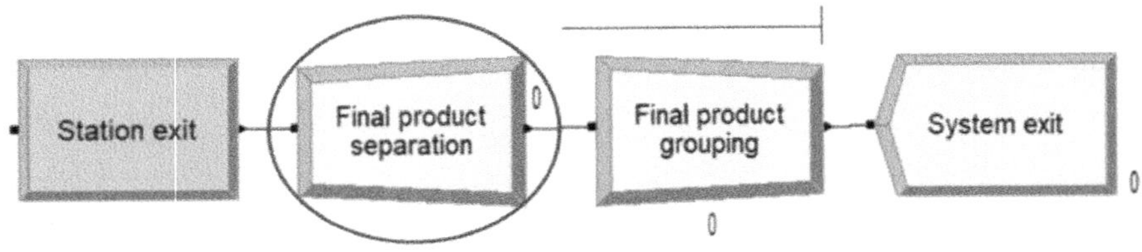

Figure 7.53. *Product release with the "nooverlapping" transfer mode. For a color version of the figure, see www.iste.co.uk/jerbi/discrete.zip*

7.2.3.2. *Cellular layout simulation model*

7.2.3.2.1. Arrival of products

This stage is identical to that used in the development of the functional layout simulation models in its two versions, each corresponding to one of the product transfer modes (Figures 7.45 and 7.46). The only difference between the models of cellular layout and functional layout lies in the sequences of product transfer. In the functional layout, products are transferred between stations corresponding to the different queues of the departments. Conversely, in the cellular layout, products are transferred between stations directly connected to the machines within the cells.

7.2.3.2.2. Cells

This stage is made up of two cells, each comprising two main components: the queue and the machines. In the cellular layout model cell adopting the "overlapping" transfer mode, when a product arrives at one of the cells, it is placed on hold in the queue of the first machine in its production sequence using a "Hold" module named "Machine Mi queue" until the machine becomes available. The selection of products from the queue follows the priority rule defined in the corresponding "Queue" module. Once the product is processed by a machine, it is transferred to the next machine in its production sequence via the "Route" module named "Route intra_cell i". This module ensures that the transfer occurs within the cells with zero transfer time. Upon completing its production sequence, each product must be discharged to the system's exit section. This is done using the "Route" module named "Rout exit cell i" with a non-zero transfer time. In cell two, sorting between products currently being processed and the products P_1, whose processing is complete, is managed through a "Decide" module named "P1 sorting" (Figure 7.54).

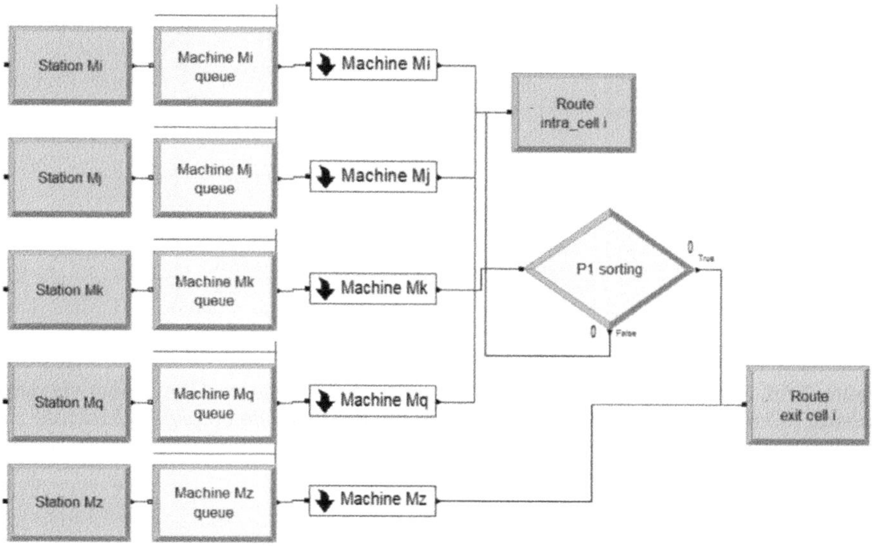

Figure 7.54. *Cell structure of the cellular model with the "overlapping" transfer mode. For a color version of the figure, see www.iste.co.uk/jerbi/discrete.zip*

In the case of the cellular model with the "nooverlapping" transfer mode, modifications are made to the cell structure at the queue level and just after the machines:

– In the cellular model with the FCFS rule, a "Separate" module, named "Separation of incoming products towards Mi", is used just before the entry of product batches into the machine queue. Once processed, the products are grouped into batches using a "Batch" module called "Grouping of outgoing products from Mi" (Figure 7.55).

– In the cellular model with the SPT rule, the queue part is represented by a "Hold" module named "Machine Mi queue". The selected batch of products then passes through a "Separate" module to separate this batch into individual products. These products are then stored in an intermediate queue represented by a "Hold" module called "Buffer hold Mi". This allows for the correct choice of the setup time reduction factor λ, based on the resemblance between the incoming products and those leaving the machine. Once processed, the products are regrouped into batches with the "Batch" module (Figure 7.56).

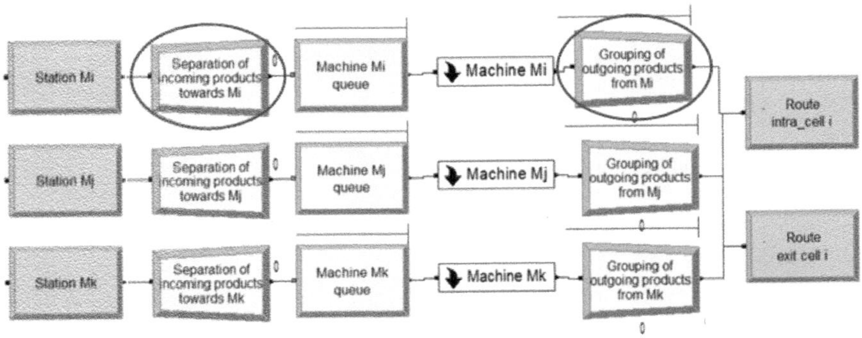

Figure 7.55. *Structure of a cellular model with the FCFS rule and the "nooverlapping" transfer mode. For a color version of the figure, see www.iste.co.uk/jerbi/discrete.zip*

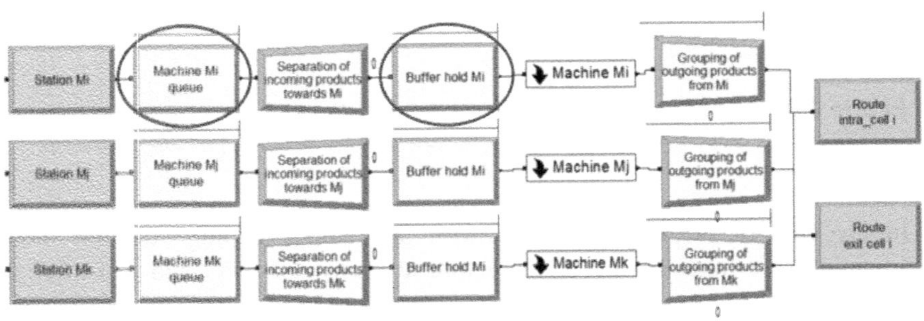

Figure 7.56. *Structure of the cellular model with the SPT rule and the "nooverlapping" transfer mode. For a color version of the figure, see www.iste.co.uk/jerbi/discrete.zip*

7.2.3.2.3. Machines

Similar to the functional layout simulation models, the machines are modeled by "Process" modules in the cellular layout. In addition, the product undergoes a test to determine the value corresponding to the setup time reduction factor λ, based on the similarity between the in-coming and out-going product type at each machine. The main distinction compared to the functional model lies in the potential values of this factor. Given that products processed within a cell belong to the same family, the value λ = 1 become unnecessary. Consequently, there are two "Assign" modules, "Lamda 0 Mij" for λ = 0 when successive products are of the same type and "Lamda 0.5 Mij" for λ = 0.5 when successive products are of different types (Figure 7.57).

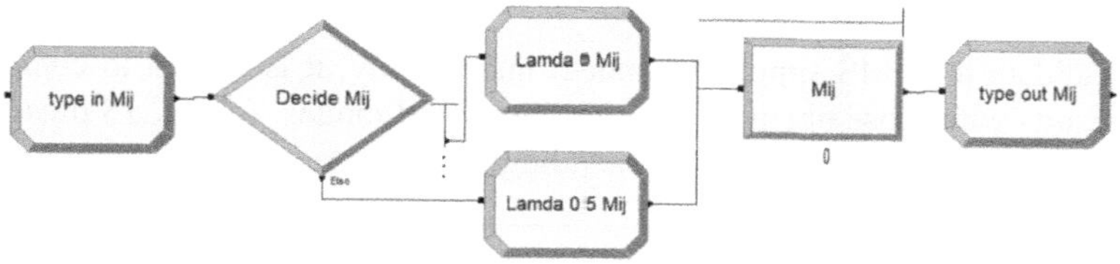

Figure 7.57. *Modeling of CL layout machines*

7.2.3.2.4. Product exit

This stage is identical to the corresponding one on functional layout simulation models (Figures 7.52 and 7.53).

7.2.4. *Verification of the simulation model*

During the development phase, the simulation model's functionality should be thoroughly reviewed and verified using animation. This process ensures that the routing of product batches is accurately verified and confirmed.

7.2.5. *Validation of the simulation model*

7.2.5.1. *Transition period*

Based on the throughput rate of the model under study, a control simulation of 100,000 minutes reveals that discarding the first 10,000 minutes is essential to mitigate the transient regime effects on the performance index of the FMS. The subsequent 90,000 minutes are then considered as the steady state (Figure 7.58).

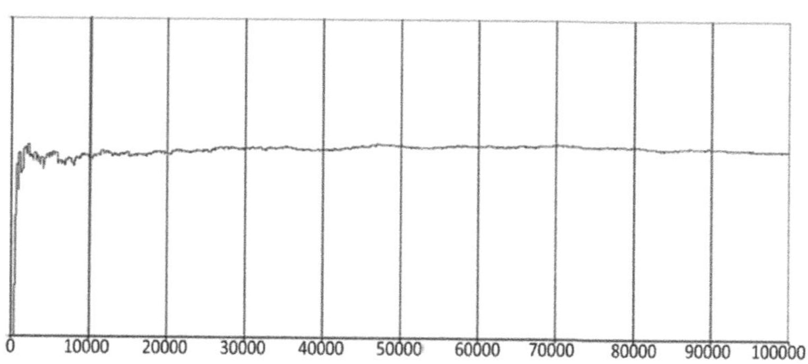

Figure 7.58. *Throughput rate of the FMS studied*

7.2.5.2. *Validation*

To validate the FMS simulation model under study, it is crucial to compare its actual results with those obtained from the simulated model. Table 7.18 presents the results of 30 simulation replications, including relevant statistical data for analysis.

Rep	Simulation	Real	Rep	Simulation	Real	Rep	Simulation	Real
1	17,003	18,002	11	17,856	18,496	21	18,101	18,102
2	18,919	17,752	12	17,000	17,685	22	17,908	16,100
3	18,209	17,402	13	17,879	18,205	23	16,973	18,052
4	17,953	17,052	14	17,818	18,307	24	17,742	18,602
5	17,619	17,921	15	17,918	17,522	25	18,734	18,710
6	17,563	17,965	16	17,516	19,102	26	18,206	17,685
7	15,964	18,052	17	17,547	18,765	27	17,656	18,007
8	17,951	17,402	18	17,200	18,702	28	19,205	17,045
9	18,513	17,422	19	18,427	17,102	29	18,643	17,586
10	18,795	17,152	20	17,586	18,952	30	18,606	18,864

Table 7.18. *MFT of simulation model and real FMS*

The graphs corresponding to these values are plotted in Figure 7.59.

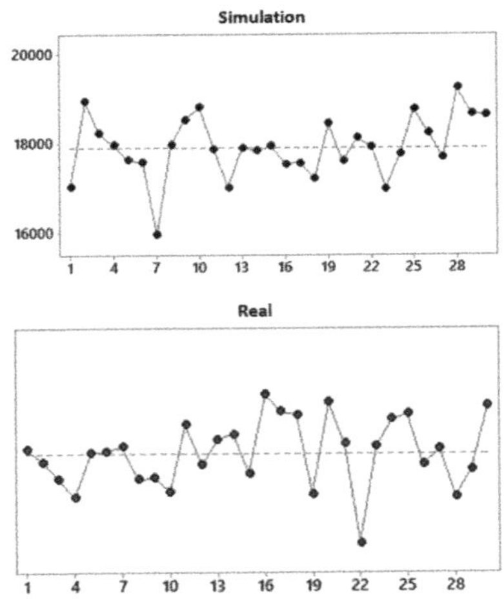

Figure 7.59. *MFT curves for the real FMS and the corresponding simulation model*

The analysis of the results enables the modeling of these values as confidence intervals, as depicted in Table 7.19.

Statistical	Simulation	Real
Average	17,900	17,924
Standard deviation	683.45	687.28
Confidence interval	[17,645; 18,156]	[17,667; 18,180]

Table 7.19. *MFT means, standard deviations and confidence intervals*

Figure 7.60 illustrates these confidence intervals, revealing that they overlap. Hence, they are considered equivalent.

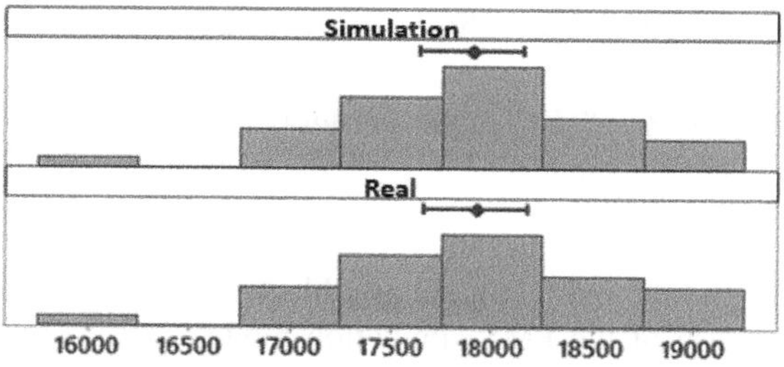

Figure 7.60. *Confidence intervals for the MFT of the real FMS and the corresponding simulation model*

A Student's *t*-test can be applied to confirm this result. This test is used at a significance level of 5% with the following hypothesis:

– H_0: difference = $N_{real} - N_{model} = 0$;

– H_1: difference = $N_{real} - N_{model} \neq 0$.

Figure 7.61 indicates that the actual difference ranges between –377.79 and 330.93. This confidence interval includes the value of 0. Moreover, the Student's *t*-test p-value is 0.895, significantly exceeding 0.05. These results suggest that there is no significant difference in MFTs between the real system and the simulation model at the 0.05 significance level. Therefore, the FMS simulation model is validated.

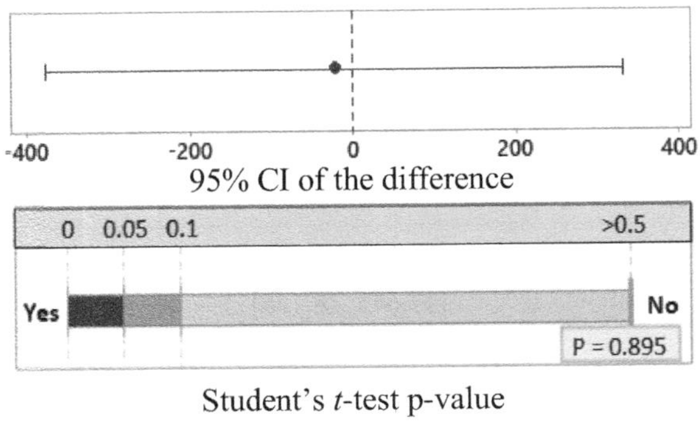

Figure 7.61. *Validation of the simulation model. For a color version of the figure, see www.iste.co.uk/jerbi/discrete.zip*

7.2.6. *Experimentation and analysis of results*

OptQuest is an optimization tool integrated into the Arena discrete event simulation software. It uses three primary search heuristics: Scatter Search (SS), Tabou Search (TS) and Neural Network (NN).

OptQuest treats the simulation model as a black box, focusing solely on its input and output. In the case of the FMS simulation model, inputs are the factor levels, while the output is the estimated performance. OptQuest automates and controls the execution of the simulation software for various FMS factor levels, initiating and concluding each simulation run and interpreting the simulation experiment results. This iterative process generates new performance values for the simulated FMS.

SS applies heuristic methods to generate an initial set of solution vectors and select the best subset as reference solutions. It then creates new points by forming a linear combination of these reference subsets. In subsequent iterations, SS selects the best solutions to initialize another round of heuristic processes. This iterative process continues until one the following three stopping criteria is reached:

– run for a specified maximum number of configurations (MNC);

– run until there is no improvement in the objective function value for 100 consecutive configurations (automatic shutdown in OptQuest);

– combination of the two rules above. In this case, the optimization algorithm is executed until the number of unimproved configurations is equal to 5% of the specified MNC.

The TS manages the reference points at each stage by using adaptive memories to move the search away from previous evaluated solutions. To speed up the search,

NNs are employed as predictive models to bypass simulations solutions that are likely inferior. The NN is trained during the search, using data gathered on the objective function values derived from various optimization variable configurations.

When applied to the FMS simulation model optimization, OptQuest begins by verifying the simulation model (Figure 7.62). The levels of the FMS factors are then adjusted according to the OptQuest algorithm. The simulation software is controlled to perform successive simulation experiments. After each experiment within the optimization cycle, the performance index value is sent from the simulation model to OptQuest for analysis, and new values for the factor levels are determined for the next iteration. This iterative process continues until the stopping criterion is reached. In our benchmarking study, OptQuest was run on a predetermined set of configurations.

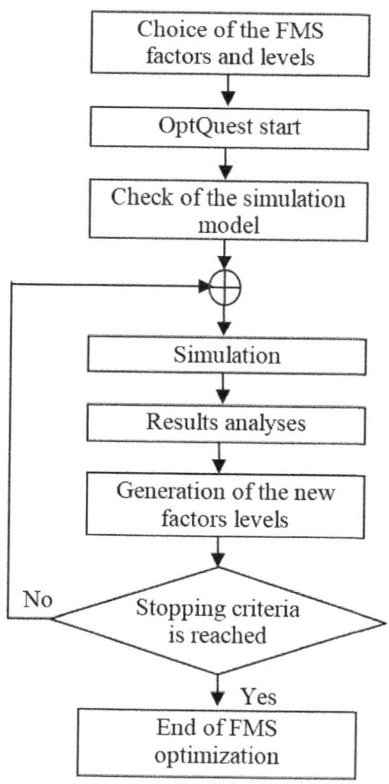

Figure 7.62. *The main steps of the simulation-based FMS optimization methodology using OptQuest*

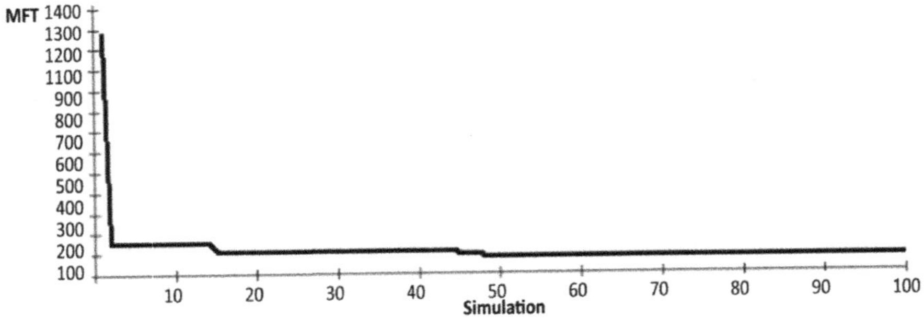

Figure 7.63. *OptQuest optimization progress for FMS MFT*

The module OptQuest was set to conduct 100 simulation experiments, with each simulation repeated three times. The optimal search rapidly converged to the minimum MFT. Indeed, during the 48th experiment, the MFT reached a minimum of 188 minutes (Figure 7.63). This optimal time was achieved with the following factor levels: BS_1, $Over_1$, PT_1, ST_1, TT_1, $layout_1$, IAT_2, $Rule_1$ and λ_1.

7.2.7. *Conclusion*

The integration of the OptQuest optimization tool with simulation proved instrumental in optimizing the average dwell time of a flexible manufacturing system. By leveraging its automated execution and sophisticated analysis capabilities, OptQuest efficiently minimized dwell time with precision, thereby achieving significant improvements in system performance. This approach not only streamlined the optimization process but also enhanced the overall operational efficiency of the manufacturing system.

8

Simulation within the Health Sector: Study of a Covid-19 Mass Vaccination Center

8.1. Introduction

In 2019, the coronavirus, better known as Covid-19, emerged and rapidly spread, becoming a global pandemic. In response, global efforts were mobilized to develop effective solutions, leading to the development of several vaccines. These vaccines aimed to stimulate the natural defenses of the immune system, enabling the development of adaptive immunity. The immune reaction to this vaccine allows the antigen against the virus to be stored so that in the event of real contamination, acquired immunity can be activated quickly and effectively.

The development of these vaccines marked the first phase of the vaccination process. Subsequently, massive vaccination programs had to be established by the health authorities of different countries around the world. These programs required meticulous planning and preparation at multiple levels, including prioritization of different categories of the population, methods of vaccine administration, public awareness, and the design and management of vaccination centers installed in appropriate locations. Tunisia established a prioritized order to implement a mass vaccination policy for its population. This prioritization considered the risk of complications, mortality by age groups and the need to maintain health and other essential professional activities. The first groups of citizens selected included individuals aged 60 years and above, followed by health professionals, essential service professionals and finally people under 60 years with specific chronic diseases. To effectively target priority citizens, Tunisian health authorities set up a national vaccination registration campaign through an IT platform. This platform assigned citizens to available vaccination centers based on appointment schedules, prioritization criteria and vaccine availability.

8.2. Objective

Effective management of vaccination centers is crucial, especially during a pandemic. Generally, governments create models without fully anticipating the results, making it necessary to verify whether these centers can achieve their established objectives. In addition, it is important to examine how these mass vaccination centers respond to the unpredictability of their operational environments.

To address these issues, simulation, and particularly discrete event simulation, can be used to evaluate the performance of a real mass vaccination center against Covid-19 in Sfax, Tunisia. This involves comparing the existing operating model of the vaccination center to a new model, providing health authorities with valuable information about the probable impact of changes in the center's layout on various performance measures. These measures include the average waiting time per citizen, the number of citizens present at the center and other performance measures under different inter-arrival time scenarios.

8.3. Description of the vaccination center

The vaccination process included several activities that could be performed by either a volunteer or a nurse depending on the nature of the task (Figure 8.1). These activities could also be performed by one or the other without any constraints. These activities were as follows:

– Registration: every citizen entering the vaccination center had to go through this point for identity and reservation verification. The operator(s) assigned to this task also checked each citizen's temperature to detect any fever or other contagious symptoms. After a rapid disinfection, the citizen was welcomed at the center with a final check to ensure they were wearing a face mask correctly. This task was usually assigned to volunteers due to its simplicity.

– Pre-vaccination check: to identify any potential contraindications to vaccination, a pre-vaccination medical assessment was conducted through a health questionnaire. This check could be carried out either by a volunteer or by a nurse using a web-based computer application.

– Vaccination: citizens who passed the pre-vaccination check moved on to the vaccination stage where one or more nurses administered the vaccine.

– Observation: after vaccination, citizens were directed to an observation area where they had to wait 15 minutes to verify the absence of adverse reactions to the vaccination.

The vaccination center studied was located in downtown Sfax within a sports complex that included both buildings and open grounds. The buildings were used for pre-vaccination checks and vaccination operations, while the grounds and adjacent driveway served as waiting and observation buffer zones, as well as a registration area. Figure 8.1 provides a summary description of the center's layout and the vaccination route that each citizen had to follow.

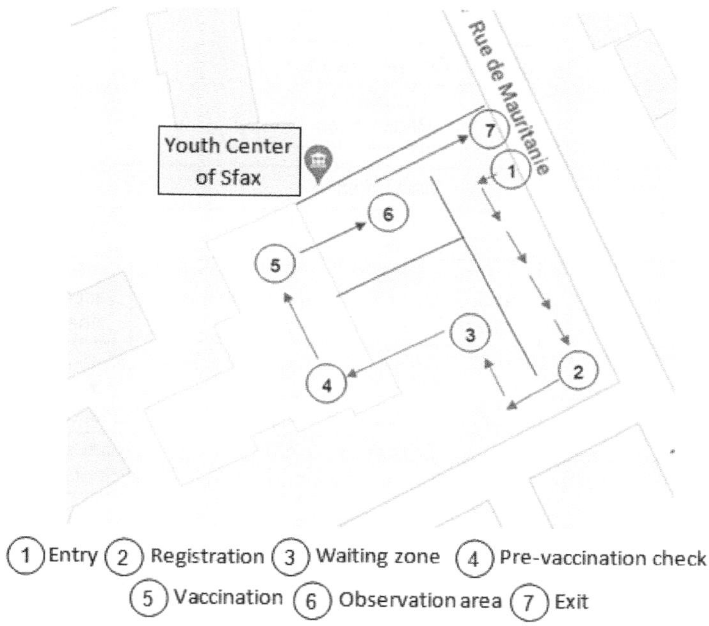

Figure 8.1. *Accessibility to the vaccination center and vaccination process. For a color version of the figure, see www.iste.co.uk/jerbi/discrete.zip*

There was a significant variability in the inter-arrival times of citizens, leading to traffic congestion on the access street for the center. In addition, the center faced challenges with citizens' non-compliance with their scheduled vaccination appointments.

8.4. Conceptual models of the vaccination center

8.4.1. *Actual model*

The registration station assigned each validated citizen a number and then directed them to a waiting area. The first six citizens in the queue were selected and directed to the first available station among the four pre-vaccination checkpoints. It was observed that approximately 2% of citizens passing through this pre-vaccination check were ineligible for vaccination. Accepted individuals were held until space became available in one of the three vaccination cubicles, each accommodating up to six individuals. Upon entering a cubicle, each citizen underwent validation to receive a confirmation SMS, proceeded to the vaccination station for the vaccine, and then moved to the observation area for medical surveillance (Figure 8.2).

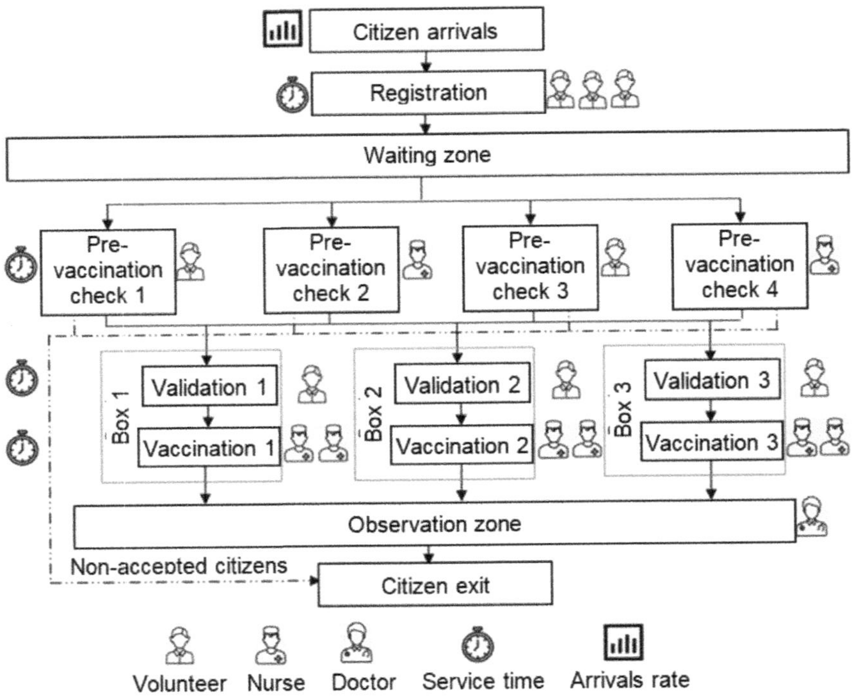

Figure 8.2. *Conceptual model of the mass vaccination center*

Three volunteers were assigned to the registration station, while two volunteers and two nurses were assigned to the four pre-vaccination checkpoints. Each vaccination cubicle was managed by a volunteer for validation operations, and two nurses for administering vaccinations. Only one doctor was assigned to the observation area, but they could be called to one of the pre-vaccination checkpoints if needed. In total, 17 staff members, including volunteers, nurses and the doctor, were necessary for the efficient operation of this vaccination center (Figure 8.2).

8.4.2. *Proposed model*

In the new model, the stages of the vaccination process are reorganized to decongest the center. Staff assignments are also reshuffled, allowing the addition of a fourth vaccination cubicle (Figure 8.3).

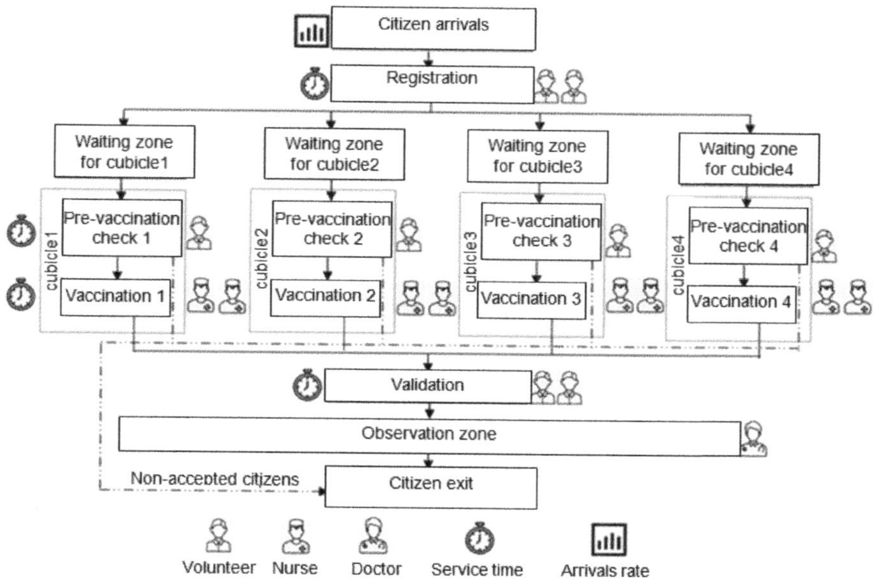

Figure 8.3. *Conceptual model of the proposed mass vaccination center*

The pre-vaccination control task is now assigned to four volunteers, and the validation task is moved outside the vaccination cubicles to reduce congestion. This validation task has been consolidated into a single station.

The two nurses previously assigned to the pre-vaccination control task are now reassigned to the new vaccination cubicle. To respect the constraint of social distancing, the waiting area is divided into four specific zones, each corresponding to one of the four vaccination cubicles. Finally, two volunteers are assigned to the registration task. The total number of operators required for the proposed model remains the same as that used in the real model (Figure 8.3).

8.5. Data collection and analysis of service times

Temporal data collection was conducted over one week at the center to gather data on service times. Several measurements of the duration of each vaccination task were collected and analyzed to fit various statistical distributions. The results of these fittings are illustrated in Figures 8.4 to 8.8. Notably, the duration of the "observation" task was constant and fixed at 15 minutes.

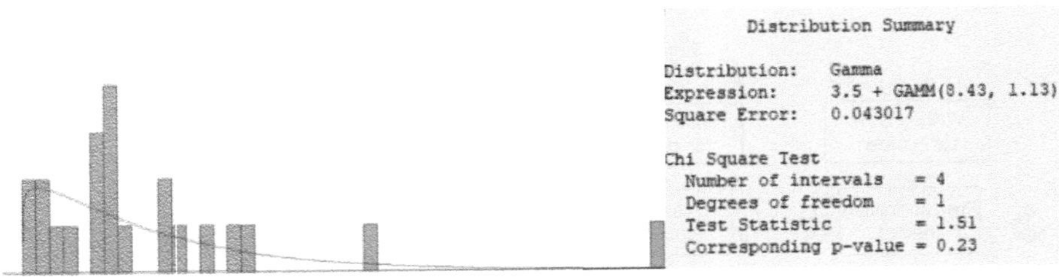

Figure 8.4. *Statistical adjustment of recording task service times*

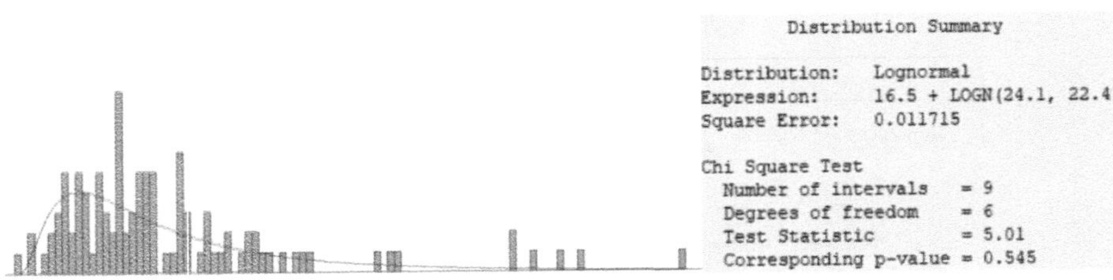

Figure 8.5. *Statistical adjustment of service times of the pre-vaccination control task*

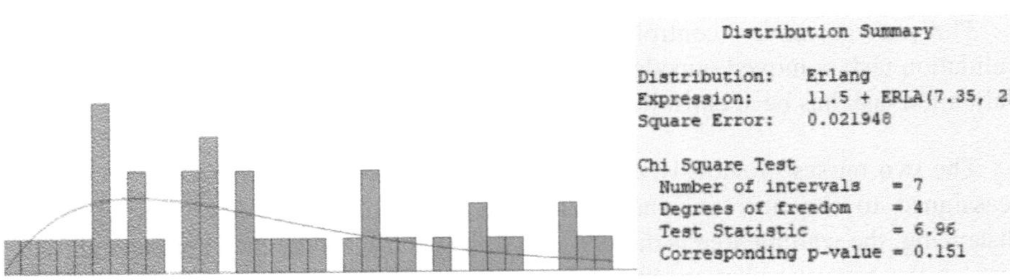

Figure 8.6. *Statistical adjustment of validation task service times*

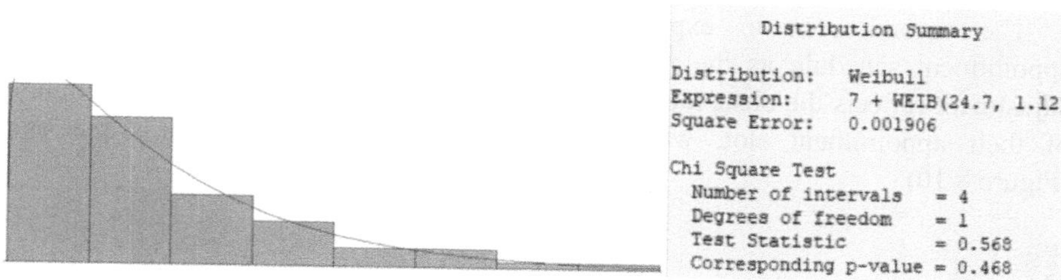

Figure 8.7. *Statistical adjustment of service times of the vaccination task: preparation and injection of the vaccine*

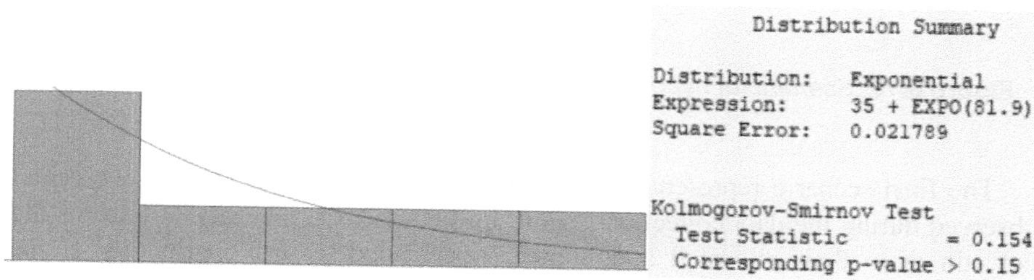

Figure 8.8. *Statistical adjustment of service times of the vaccination task: recovery of the vial from the pharmacy*

The adequacy tests show that all statistical distributions adequately model the collected service time data.

8.6. Arrival rate of citizens at the center

The arrivals of citizens at the vaccination center followed a theoretical scenario established by the Tunisian health authorities. In this scenario, appointments were scheduled for 90 citizens per half hour from 8:30 a.m. to 5 p.m. (Figure 8.9).

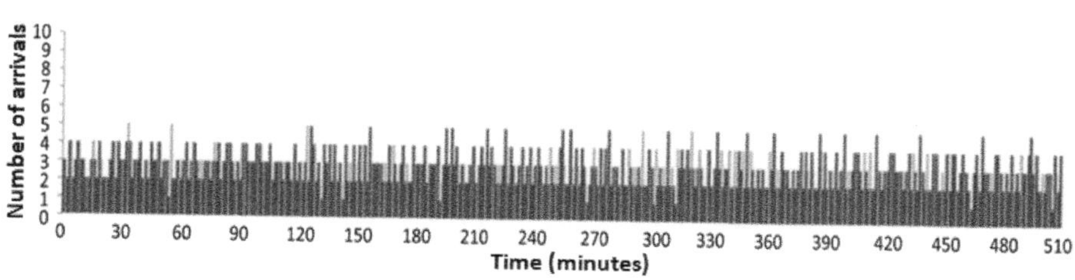

Figure 8.9. *Scenario of theoretical arrivals of citizens at the vaccination center*

The second scenario explores hypothetical arrivals based on the same appointment schedule as the first scenario but incorporates random noise. This adjustment reflects the observation that most citizens tend to arrive at the beginning of their appointment slot, while a small percentage arrive towards the end (Figure 8.10).

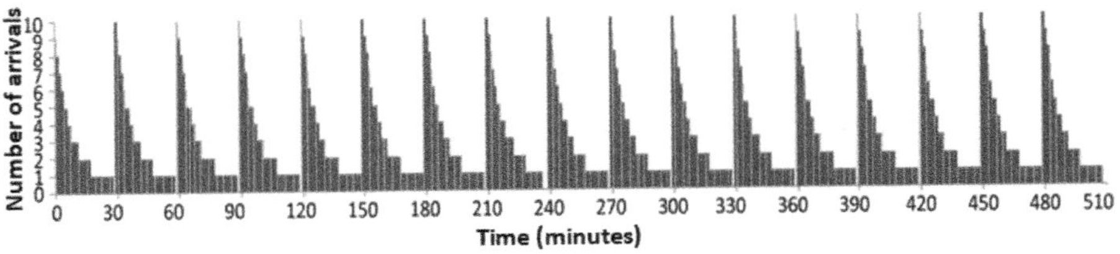

Figure 8.10. *Scenario of hypothetical arrivals of citizens at the vaccination center*

The final scenario represents the actual arrival patterns of citizens at the center observed during the data collection period. In this case, the majority of citizens did not adhere to their allocated appointment and tended to arrive primarily at the beginning and in the middle of the day (Figure 8.11).

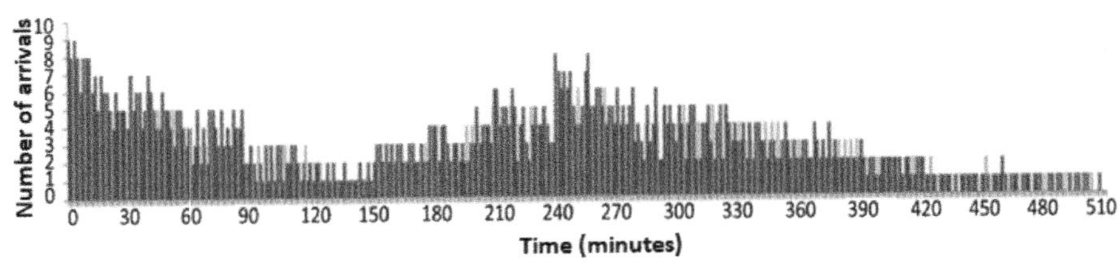

Figure 8.11. *Scenario of real arrivals of citizens at the vaccination center*

All these arrival scenarios were programmed through the "Schedule" feature, allowing the arrival times of citizens to be programmed according to a schedule.

8.7. Coding the simulation model

8.7.1. *Actual model*

8.7.1.1. *Entry and registration*

A "Create" module, called "Citizen entry", serves as a generation point for citizens arriving at the simulation model. The entities representing the citizens leave this module and pass through an "Assign" module, named "Input attributes", to increment several attributes necessary for calculating system performance. These entities then go through a "Process" module, named "Registration and allocation of a number", which holds them for the time required for registration and number allocation. This time follows the statistical distribution modeled for this task, as shown in Figure 8.4.

Next, the entity representing the citizen is directed to the "Hold" module, named "Waiting zone", which acts as a queue for the pre-vaccination checkpoints. Once a position becomes available, the first six citizens are released and grouped by the "Batch" module named "Checkpoint grouping". This group is then sent to the available station through the "PickStation" module, named "Selection pre-vaccination control station" (Figure 8.12).

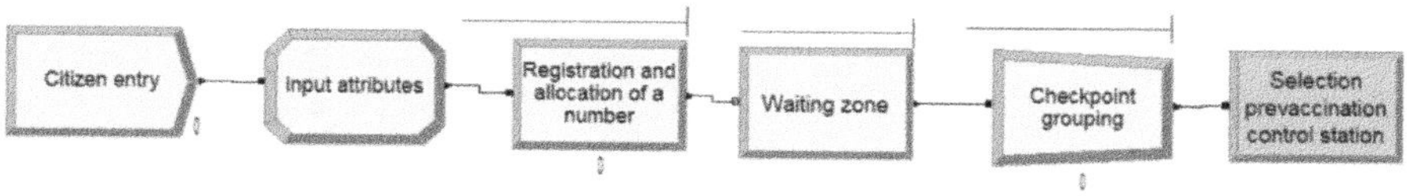

Figure 8.12. *Entry, registration and numbering area (actual model). For a color version of the figure, see www.iste.co.uk/jerbi/discrete.zip*

8.7.1.2. *Pre-vaccination control*

The group of six citizens entering one of the pre-vaccination control zones is separated by the "Separate" module, named "Degrouping prevaccination control i". These six citizens wait in a common line at the two checkpoints, modeled by a "Hold" module named "Queue prevaccination control i". The checks are performed cyclically by the staff assigned to these positions, with the control stations modeled by "Process" modules. These modules retain the entity representing the citizen for the pre-vaccination control period, following the statistical distribution modeled for this task. Upon exiting the control, the entities pass through a "Decide" module, which removes 2% of citizens considered unsuitable for vaccination. Eligible citizens are placed in an intermediate queue, modeled by a "Hold" module named "Waiting for vaccination", until one of the three vaccination stations becomes available. The choice between these stations is made through a "PickStation" module, named "Choice of vaccination station", based on the number of pending "Process" modules representing the vaccination stations (Figure 8.13).

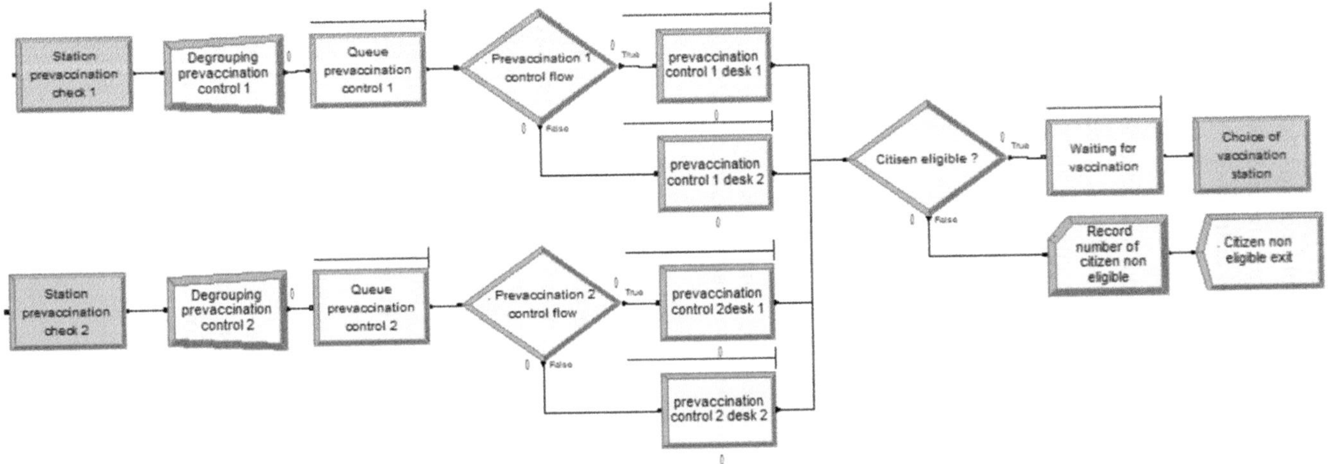

Figure 8.13. *Pre-vaccination control zone (real model). For a color version of the figure, see www.iste.co.uk/jerbi/discrete.zip*

8.7.1.3. *Vaccination and observation*

Upon entering one of the vaccination cubicles, each entity representing a citizen undergoes two stages: validation and vaccination. Two "Process" modules represent these stations. The first module sends a vaccination SMS to the citizen via a computer application, while the second module simulates the process of administering the vaccine.

The vaccination station modules each include two resources representing the nurses assigned to this task. The first nurse prepares the dose of vaccine, while the second inoculates it for the citizen. This position retains the entity representing the citizen for the duration of time modeled by the statistical distribution relating to the vaccination task encompassing both vaccine preparation and injection. These modules pause whenever the first nurse, responsible for preparing the vaccine doses, needs to replenish supplies as indicated by the downtime, which is modeled by the statistical distribution given in Figure 8.8.

After vaccination, each citizen proceeds to an observation zone modeled by a "Hold" module named "Observation zone", where they remain for a period of 15 minutes. Upon completion of the observation period, these entities pass through an "Assign" module named "Output attributes", which updates several attributes necessary for calculating system performance. These performance metrics are collected by the "Record" module named "Output recording" each time an entity passes through. Finally, these entities exit the simulation model through the "Dispose" module named "Citizen exit" (Figure 8.14).

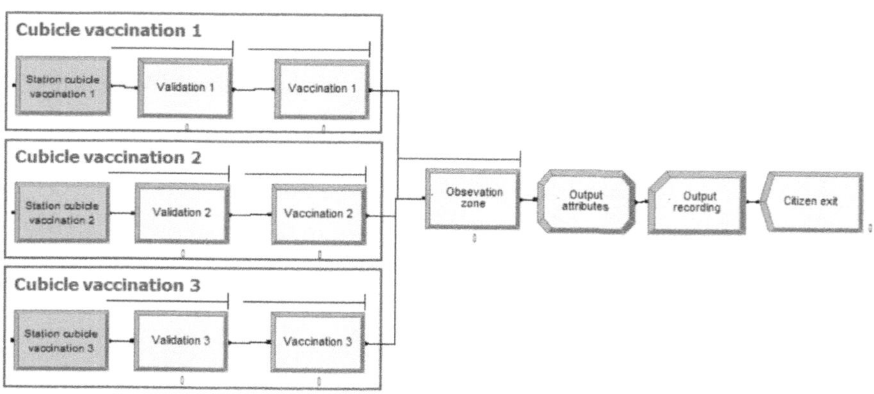

Figure 8.14. *Vaccination area and post-vaccination observation (real model). For a color version of the figure, see www.iste.co.uk/jerbi/discrete.zip*

8.7.2. *Proposed vaccination center*

8.7.2.1. *Entrance and registration*

Entities are generated by the "Create" module, named "Citizen entry", and then proceed to the "Assign" module, named "Entry attributes". Following this, they go through the "Process" module, called "Registration and number allocation". In contrast to the real center model, the "Hold" and "Batch" modules, named "Waiting zone" and "Checkpoint grouping", are no longer necessary and are removed from this zone. Each registered citizen is subsequently directed to the free or least busy queue of the four vaccination cubicles through the "PickStation" module named "Choice of vaccination station" (Figure 8.15).

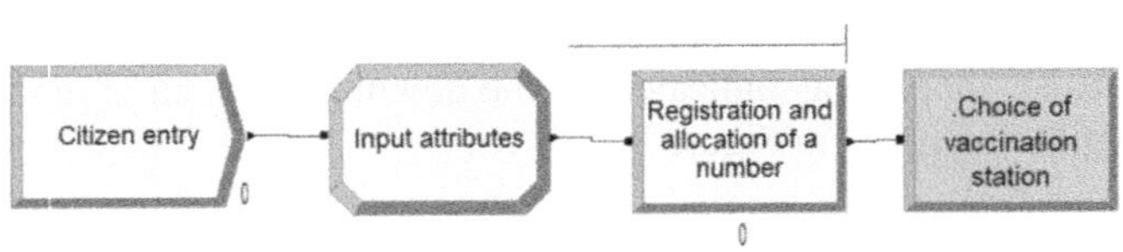

Figure 8.15. *Entry, registration and numbering area (proposed center). For a color version of the figure, see www.iste.co.uk/jerbi/discrete.zip*

8.7.2.2. *Pre-vaccination control and vaccination*

Each citizen is put on hold in a "Hold" module named "Waiting zone for cubicle i" while awaiting availability at the corresponding pre-vaccination checkpoint. Upon exiting the checkpoint, citizens go through a "Decide" module, named "Citizen eligible cubicle i?", where 2% of citizens not suitable for vaccination are removed. The remaining citizens proceed to the "Process" module named "Vaccination i", which simulates the vaccinating process (Figure 8.16).

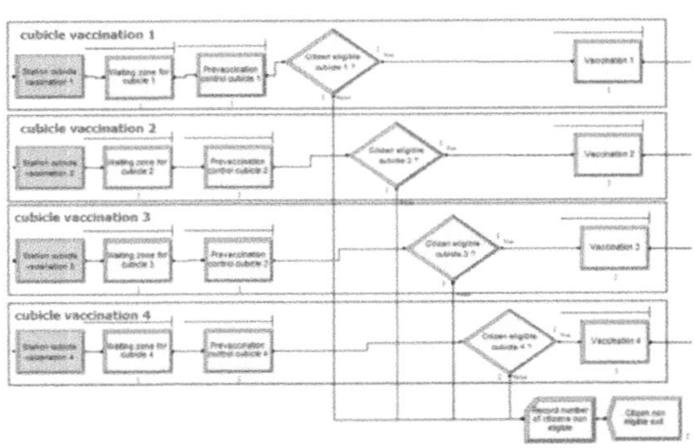

Figure 8.16. *Pre-vaccination control zone and vaccination (proposed model). For a color version of the figure, see www.iste.co.uk/jerbi/discrete.zip*

8.7.2.3. *Validation and post-vaccination observation*

After vaccination, each citizen proceeds to the single validation station where they receive a vaccination SMS. Subsequently, they pass through the same “Hold”, “Assign” and “Record” modules corresponding to this area of the simulation model of the real center. Finally, citizens exit the simulation model through the “Dispose” module named “Citizen exit” (Figure 8.17).

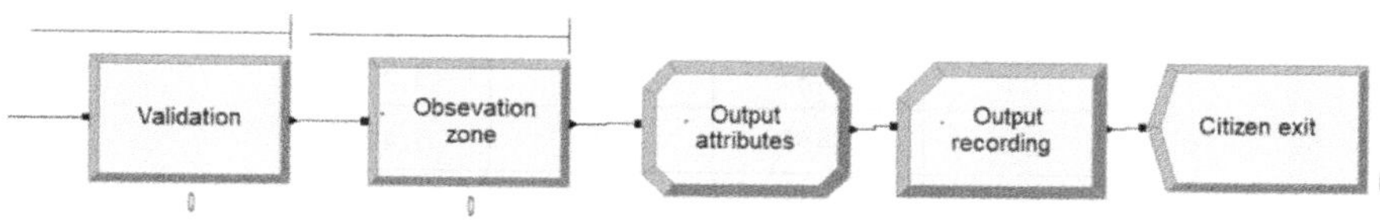

Figure 8.17. *Post-vaccination validation and observation zone (proposed model)*

8.8. Verification of simulation models

The simulation models of the vaccination center underwent verification both during and at the end of the coding phase using the Arena software. This verification was conducted in the presence of vaccination center management personnel through the animation of the flow of entities during the simulation. This mainly focused on verifying the concordance between the movement of entities representing citizens in the simulation model and the operational prescriptions of the conceptual model.

To validate the simulation, stochastic service times were initially substituted with constant times. This adjustment aimed to verify the concordance between the entities’ entries and exits in the different stages of the vaccination process. In addition, these times and the number of arrivals were systematically also varied towards extreme values to verify the robustness of the simulation models against different scenarios encountered in the vaccination center’s operations.

8.9. Validation of the simulation model of the real center

8.9.1. *Simulation replication length*

The vaccination center functions as a terminating system, operating continuously for 510 minutes each day before shutting down. Each operational period begins independently without any influence from the previous period. For simulation purposes, each replication was set to run for 510 minutes, and after completing each replication, all statistics were reset and the simulation model started in an empty state.

8.9.2. Calculating the number of simulation replications

Ten simulation replications of the real model were performed using the real arrivals scenario to determine the final number of simulation replications. During these replications, data were collected on the number of vaccinated citizens (Table 8.1).

Replication	NOUT	Replication	NOUT
1	1,487	6	1,573
2	1,468	7	1,461
3	1,446	8	1,549
4	1,509	9	1,423
5	1,490	10	1,483

Table 8.1. *Result of the 10 simulation replications, number of vaccinated citizens*

The average number of vaccinated citizens was 1488.8, with a standard deviation of 45.37. Calculating the confidence interval gives CI = 32.42 with $t_{1-0.05/2,\ 10-1} = 2.26$. For an absolute precision of 19 citizens, the required number of final replications was calculated as n = 29.12, which can be rounded to 30, with summarized results presented in Table 8.2.

Replication	NOUT	Replication	NOUT	Replication	NOUT
1	1,487	11	1,493	21	1,482
2	1,468	12	1,424	22	1,464
3	1,446	13	1,495	23	1,525
4	1,509	14	1,462	24	1,504
5	1,490	15	1,460	25	1,399
6	1,573	16	1,508	26	1,570
7	1,461	17	1,389	27	1,540
8	1,549	18	1,456	28	1,413
9	1,423	19	1,431	29	1,432
10	1,483	20	1,527	30	1,467

Table 8.2. *Result of the final 30 simulation replications, number of vaccinated citizens*

To ensure that these 30 replications are sufficient for the desired level of absolute precision, the mean and standard deviation were recalculated to be 1,477.67 and 47.61 citizens, respectively. The recalculated confidence interval is now CI = 17.73 citizens with $t_{1-0.05/2,\,30-1} = 2.05$. This CI value is lower than the chosen level of precision, indicating that the number of replications was indeed sufficient.

8.9.3. *Validation*

To validate the simulation model, the results obtained from the simulation were compared with actual statistics from the vaccination center. Table 8.3 presents the outcomes of 30 simulation replications of the vaccination center using the real arrival scenario, alongside the vaccination statistics observed at the real center over the duration of one month.

Rep	$NOUT_{center}$	$NOUT_{model}$	Rep	$NOUT_{center}$	$NOUT_{model}$	Rep	$NOUT_{center}$	$NOUT_{model}$
1	1,480	1,487	11	1,455	1,493	21	1,458	1,482
2	1,420	1,468	12	1,502	1,424	22	1,465	1,464
3	1,414	1,446	13	1,507	1,495	23	1,485	1,525
4	1,490	1,509	14	1,402	1,462	24	1,485	1,504
5	1,520	1,490	15	1,450	1,460	25	1,502	1,399
6	1,524	1,573	16	1,482	1,508	26	1,558	1,570
7	1,432	1,461	17	1,444	1,389	27	1,562	1,540
8	1,522	1,549	18	1,559	1,456	28	1,432	1,413
9	1,480	1,423	19	1,504	1,431	29	1,452	1,432
10	1,432	1,483	20	1,408	1,527	30	1,600	1,467

Table 8.3. *Result of the numbers of citizens vaccinated by the vaccination center and by the corresponding simulation model*

The corresponding curves for these values are illustrated in Figure 8.18. The analysis of the results from these 30 replications enabled modeling them with confidence intervals, as depicted in Table 8.4.

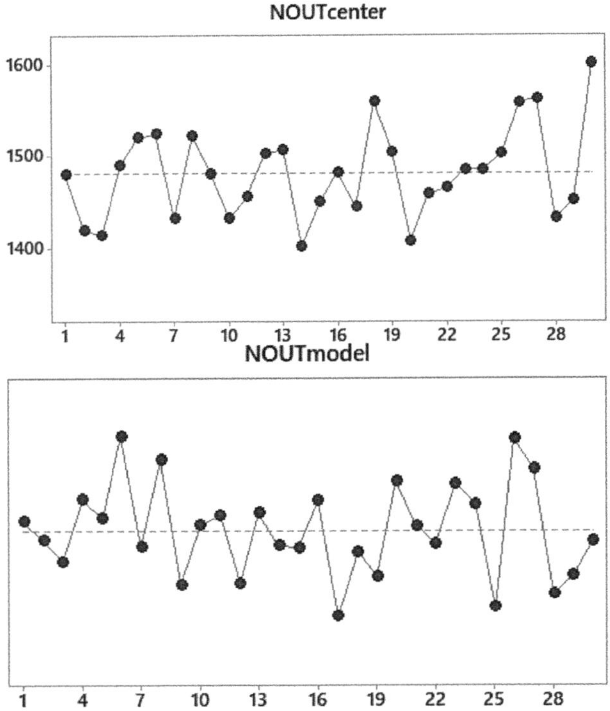

Figure 8.18. *Curves of the numbers of citizens vaccinated by the vaccination center and by the corresponding simulation model*

Statistical	$NOUT_{center}$	$NOUT_{model}$
Average	1480.86	1477.7
Standard deviation	49,689	47,614
Confidence interval	[1462,359; 1499.373]	[1459.933; 1495,400]

Table 8.4. *Means, standard deviations and confidence intervals of the number of citizens vaccinated by the center and by the corresponding simulation model*

The confidence intervals for the numbers of citizens vaccinated by the real center and the simulation model, as shown in Figure 8.19, overlap. This indicates that these results are equivalent.

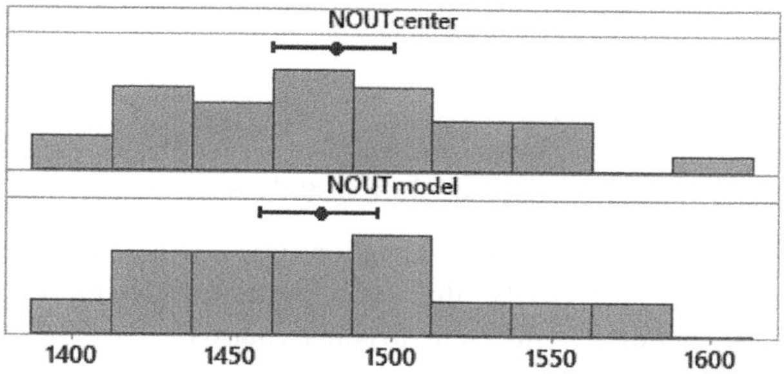

Figure 8.19. *Confidence intervals of the numbers of citizens vaccinated by the center and by the corresponding simulation model*

A Student's *t*-test was applied to confirm the results of the comparison. This *t*-test was used at a significance level of 5% with the following hypothesis:

– H_0: difference = $NOUT_{center} - NOUT_{model} = 0$;

– H_1: difference = $NOUT_{center} - NOUT_{model} \neq 0$.

The confidence interval of the difference between $NOUT_{center}$ and $NOUT_{model}$ quantifies the uncertainty associated with estimating the difference of means from sample data. As shown in Figure 8.20, there is 95% confidence that the real difference is between –17.726 and 24.126. The value 0 falls within this confidence interval and the p-value of the Student's *t*-test is 0.8. These results thus reveal that the means are equivalent to the significance level of 0.05. Therefore, the simulation model of the real center can be considered valid.

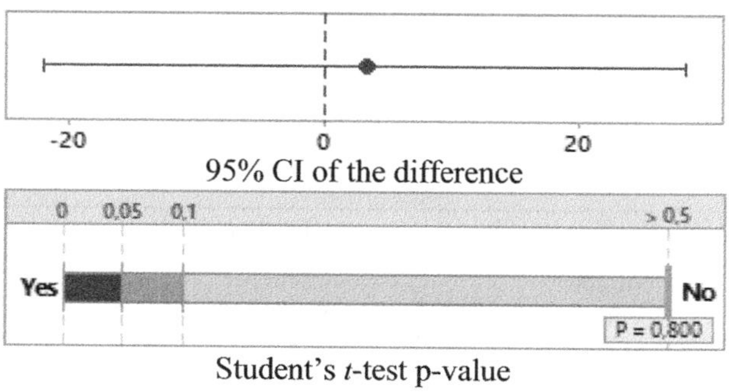

Figure 8.20. *Validation of the simulation model. For a color version of the figure, see www.iste.co.uk/jerbi/discrete.zip*

8.10. Experimentation and analysis of results

8.10.1. *Performance indicators*

The performance indicators adopted for the study were as follows:

– The total number of citizens vaccinated on time compared to the total number of citizens arrived (NOUT/NAR): this ratio evaluates the effectiveness of the vaccination center according to different arrival scenarios.

– The average total vaccination time per citizen (TVT): this time characterizes the fluidity of the citizens' flow through the center. A lower TVT indicates a more efficient and fluid process.

– Waiting time per citizen (WT): this is the total time a citizen spends waiting at each stage of the vaccination process. Minimizing this time improves the overall efficiency and citizen experience.

– The number of citizens in the center (NCI): this measure ensures that the social distancing condition necessary to limit the spread of Covid-19 is met. A higher NCI makes it more difficult to ensure this condition.

– The average operator utilization rate (UTIL): this is calculated by dividing the total time spent by each operator by the length of the working day. Maximizing, this average improves operator performance and resource utilization.

8.10.2. *Analysis of simulation results*

8.10.2.1. *Comparison on the NOUT/NAR basis*

The simulation results for the number of arrivals and number out of the two vaccination center models for the three arrival scenarios are summarized in Table 8.5.

Rep	Theoretical arrivals			Hypothetical arrivals			Actual arrivals		
	Number of arrivals	Number vaccinated		Number of arrivals	Number vaccinated		Number of arrivals	Number vaccinated	
		Real	Proposed		Real	Proposed		Real	Proposed
1	1,585	1,487	1,485	1,575	1,494	1,525	1,523	1,487	1,478
2	1,520	1,424	1,429	1,583	1,493	1,513	1,520	1,468	1,481
3	1,490	1,403	1,405	1,508	1,452	1,447	1,493	1,446	1,442
4	1,581	1,463	1,473	1,521	1,403	1,428	1,546	1,509	1,517
5	1,534	1,451	1,471	1,569	1,509	1,518	1,523	1,490	1,487
6	1,562	1,480	1,481	1,577	1,523	1,524	1,611	1,573	1,569

Rep	Theoretical arrivals			Hypothetical arrivals			Actual arrivals		
	Number of arrivals	Number vaccinated		Number of arrivals	Number vaccinated		Number of arrivals	Number vaccinated	
		Real	Proposed		Real	Proposed		Real	Proposed
7	1,623	1,533	1,547	1,531	1,464	1,475	1,503	1,461	1,469
8	1,574	1,483	1,487	1,492	1,410	1,438	1,587	1,549	1,547
9	1,496	1,387	1,388	1,573	1,474	1,502	1,459	1,423	1,422
10	1,485	1,386	1,416	1,470	1,397	1,409	1,520	1,483	1,479
11	1,574	1,486	1,492	1,532	1,434	1,479	1,531	1,493	1,500
12	1,563	1,469	1,477	1,538	1,473	1,488	1,462	1,424	1,429
13	1,555	1,468	1,477	1,517	1,448	1,459	1,530	1,495	1,492
14	1,562	1,467	1,481	1,564	1,476	1,487	1,502	1,462	1,473
15	1,633	1,536	1,543	1,497	1,431	1,451	1,503	1,460	1,461
16	1,539	1,450	1,467	1,538	1,460	1,470	1,550	1,508	1,512
17	1,488	1,399	1,411	1,567	1,491	1,502	1,419	1,389	1,392
18	1,548	1,451	1,466	1,542	1,459	1,476	1,491	1,456	1,454
19	1,555	1,453	1,462	1,530	1,458	1,472	1,468	1,431	1,427
20	1,536	1,452	1,454	1,485	1,414	1,417	1,573	1,527	1,534
21	1,564	1,462	1,470	1,536	1,457	1,463	1,520	1,482	1,476
22	1,536	1,431	1,463	1,633	1,560	1,571	1,509	1,464	1,477
23	1,528	1,428	1,442	1,487	1,397	1,429	1,573	1,525	1,522
24	1,537	1,449	1,457	1,517	1,443	1,451	1,533	1,504	1,502
25	1,493	1,404	1,415	1,537	1,467	1,475	1,438	1,399	1,401
26	1,542	1,460	1,471	1,566	1,487	1,496	1,613	1,570	1,564
27	1,547	1,455	1,456	1,557	1,459	1,496	1,584	1,540	1,546
28	1,489	1,395	1,398	1,538	1,449	1,448	1,464	1,413	1,425
29	1,480	1,403	1,407	1,519	1,442	1,454	1,479	1,432	1,438
30	1,507	1,413	1,416	1,563	1,483	1,493	1,512	1,467	1,482

Table 8.5. *Simulation results for NOUT and NAR (in citizens)*

The 95% confidence intervals for the NOUT/NAR ratios of the two simulation models under the three arrival scenarios are illustrated in Figure 8.21. The average rates of citizens receiving the vaccine on time in the two vaccination center models indicate that most citizens receive their vaccines on schedule. The proposed center model outperforms the real center in the theoretical and hypothetical arrival scenarios. However, the real arrivals scenario does not show significant differences between the two simulation models.

The confidence intervals of the difference between the models across the different scenarios as well as the pairwise Student's t-tests confirm this result (Figure 8.22). The t-tests yield p-values less than 0.001 for both the theoretical and hypothetical arrival scenarios, indicating significant differences between the real and proposed center models. Conversely, the p-value for the real arrivals scenario is 0.066, suggesting no significant difference between the models in this scenario (Figure 8.22).

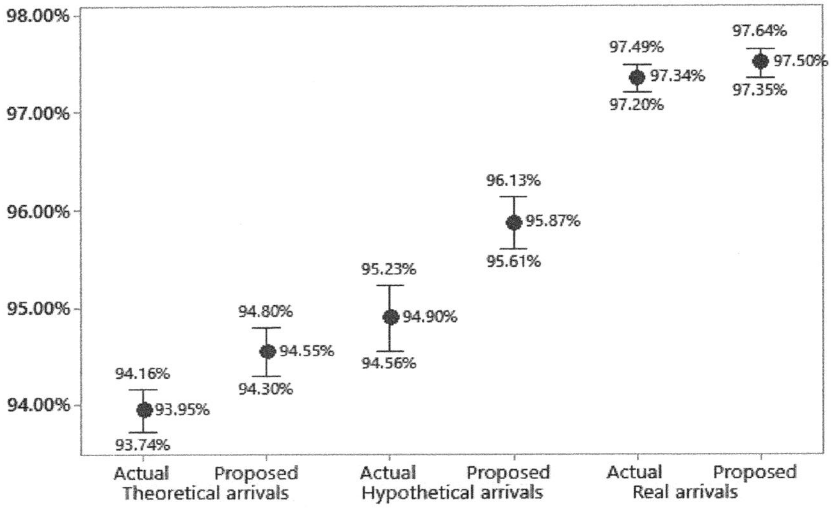

Figure 8.21. *The 95% CI for the ANOUT/ANAR ratio (in %)*

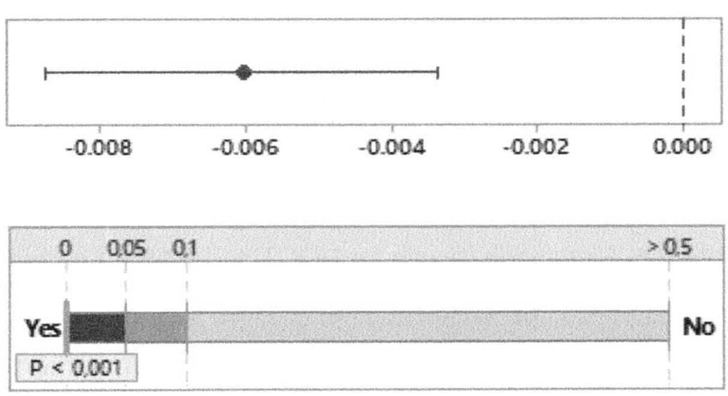

a) Theoretical arrivals scenario

Figure 8.22A. *Pairwise comparison of NOUT/NAR ratios (in %) of the real and proposed models. For a color version of the figure, see www.iste.co.uk/jerbi/discrete.zip*

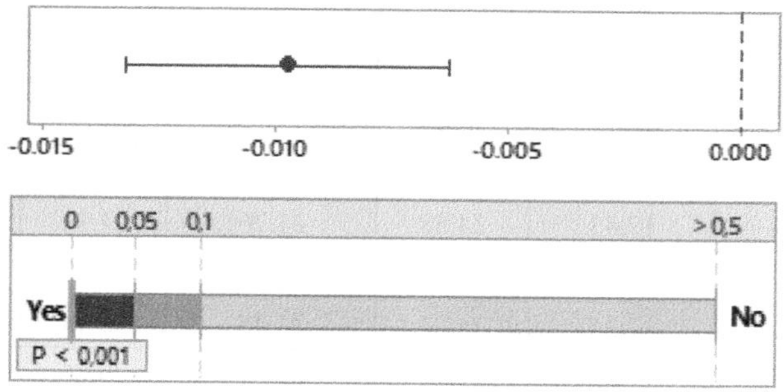

b) Hypothetical arrivals scenario

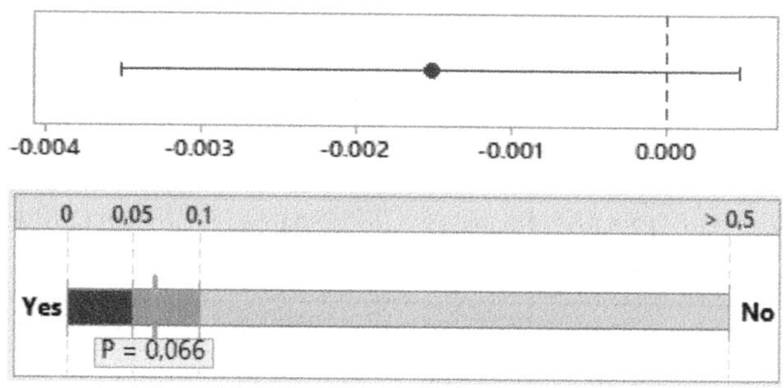

c) Actual arrivals scenario

Figure 8.22B. *Pairwise comparison of NOUT/NAR ratios (in %) of actual and proposed models (continued). For a color version of the figure, see www.iste.co.uk/jerbi/discrete.zip*

8.10.2.2. *Comparison based on TVT*

Table 8.6 gives the simulation results for the TVT per citizen.

Rep	Theoretical arrivals		Hypothetical arrivals		Actual arrivals	
	Actual model	Proposed model	Actual model	Proposed model	Actual model	Proposed model
1	19.64	17.34	24.09	21.45	35.82	22.22
2	19.58	17.35	23.71	21.28	36.45	23.49
3	19.57	17.20	24.35	21.65	32.90	21.00
4	19.74	17.61	23.06	21.06	36.59	22.07
5	20.08	17.43	25.02	21.19	31.81	21.42
6	19.98	17.33	24.21	21.61	40.52	22.49
7	19.59	17.53	23.40	21.26	28.72	21.36

Rep	Theoretical arrivals		Hypothetical arrivals		Actual arrivals	
	Actual model	Proposed model	Actual model	Proposed model	Actual model	Proposed model
8	19.57	17.47	23.40	20.90	34.07	21.22
9	19.48	17.43	24.58	21.76	34.11	21.48
10	19.70	17.46	23.38	20.99	33.92	22.21
11	19.93	17.36	24.25	21.45	35.14	21.04
12	19.78	17.53	23.70	21.11	29.86	19.30
13	20.36	17.46	24.86	21.22	40.66	21.35
14	19.87	17.54	23.75	21.10	29.92	19.87
15	20.00	17.55	23.50	21.14	32.49	21.90
16	19.72	17.31	23.77	20.72	33.21	21.98
17	19.43	17.27	23.99	21.52	27.49	19.91
18	19.59	17.57	24.18	21.14	30.02	22.12
19	20.09	17.33	25.03	21.35	33.12	22.04
20	19.90	17.46	23.45	20.69	38.59	24.00
21	19.57	17.30	23.66	20.59	35.15	20.90
22	19.65	17.45	24.62	21.55	32.42	21.95
23	19.33	17.27	23.03	20.62	35.65	22.60
24	20.11	17.35	24.61	21.14	31.00	21.27
25	19.54	17.29	24.41	21.07	35.10	20.49
26	19.75	17.55	23.73	21.29	39.67	23.10
27	19.83	17.28	24.06	21.44	37.20	24.25
28	19.84	17.30	24.33	21.22	30.64	22.15
29	19.58	17.40	23.59	20.73	33.67	20.82
30	19.64	17.30	24.36	21.41	29.30	20.97

Table 8.6. *TVT simulation results (in minutes)*

Figure 8.23 presents the 95% confidence intervals for the TVT across all simulation models. In the real model, the TVT increases from 19.748 minutes for theoretical arrivals to 24.002 minutes for hypothetical arrivals and 33.839 minutes for real arrivals. Conversely, the proposed model maintains comparable values across all scenarios, with averages of 17.401 minutes, 21.189 minutes and 21.698 minutes, respectively. Furthermore, there is no overlap between the CIs for the TVTs of the two models, which is a significant difference. This is further confirmed by the CIs of the differences between the two models and the results of the Student's *t*-tests (Figure 8.24). None of the CIs include the value of 0, and the p-values of the tests are all less than 0.001, underscoring the significant difference between the real and proposed simulation models.

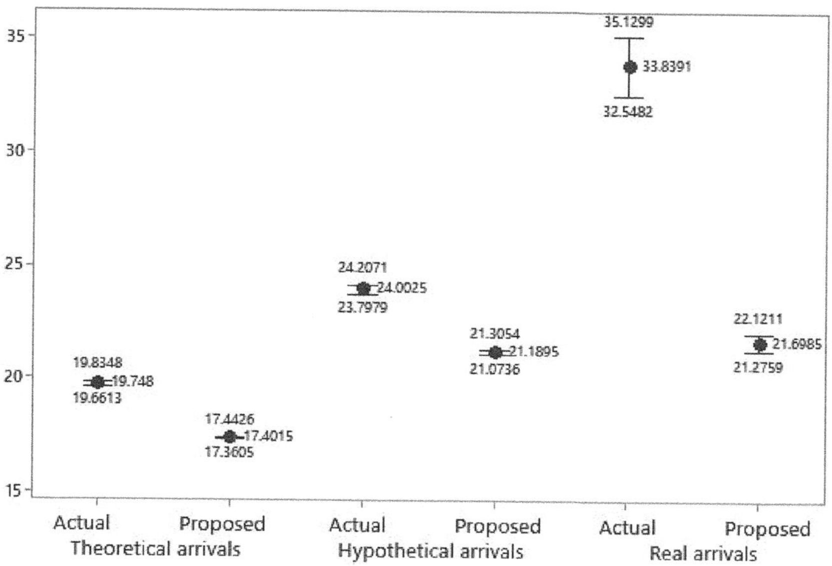

Figure 8.23. *The 95% CI for TVT (in minutes)*

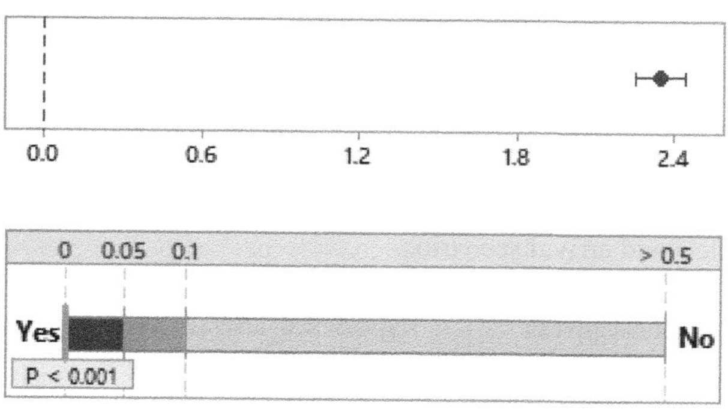

a) Theoretical arrivals scenario

Figure 8.24A. *Pairwise comparison of TVT (in minutes) of real and proposed models. For a color version of the figure, see www.iste.co.uk/jerbi/discrete.zip*

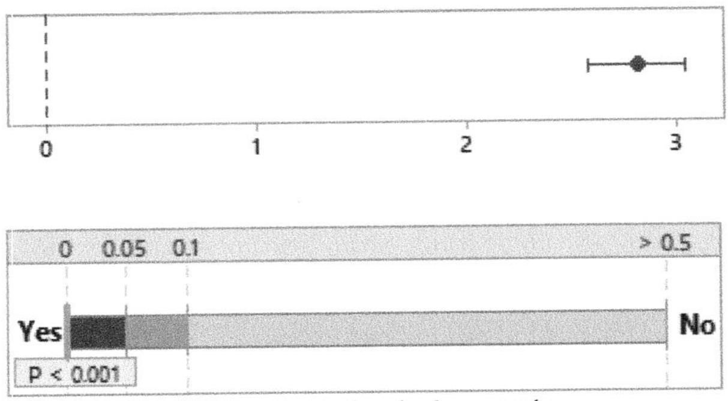

b) Hypothetical arrivals scenario

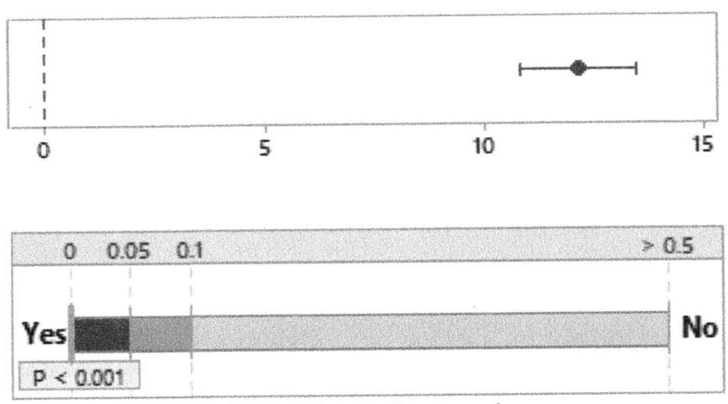

c) Actual arrivals scenario

Figure 8.24B. *Pairwise comparison of TVT (in minutes) of the actual and proposed models (continued). For a color version of the figure, see www.iste.co.uk/jerbi/discrete.zip*

8.10.2.3. *Comparison based on WT*

Table 8.7 gives the simulation results of the average WT per citizen for the two simulation models and arrival scenarios.

Rep	Theoretical arrivals		Hypothetical arrivals		Actual arrivals	
	Actual model	Proposed model	Actual model	Proposed model	Actual model	Proposed model
1	3.09	0.86	7.67	4.88	19.23	5.74
2	3.12	0.90	7.34	4.78	20.00	6.96
3	3.01	0.73	7.72	5.07	16.40	4.60
4	3.31	1.12	6.61	4.42	20.07	5.48

Rep	Theoretical arrivals		Hypothetical arrivals		Actual arrivals	
	Actual model	Proposed model	Actual model	Proposed model	Actual model	Proposed model
5	3.53	0.82	8.43	4.74	15.22	4.87
6	3.48	0.84	7.64	5.11	23.97	6.03
7	3.20	1.02	6.93	4.73	12.23	4.81
8	3.02	0.90	6.94	4.38	17.54	4.72
9	3.03	0.95	8.03	5.20	17.57	4.98
10	3.16	0.81	6.84	4.39	17.37	5.73
11	3.47	0.83	7.83	4.87	18.56	4.44
12	3.21	0.97	7.15	4.59	13.31	2.80
13	3.88	0.93	8.28	4.66	24.08	4.87
14	3.33	0.94	7.26	4.66	13.42	3.29
15	3.56	1.09	6.98	4.57	16.02	5.43
16	3.29	0.81	7.22	4.26	16.74	5.45
17	2.91	0.70	7.46	4.97	10.87	3.26
18	3.06	0.93	7.64	4.60	13.48	5.61
19	3.61	0.83	8.50	4.74	16.57	5.56
20	3.38	0.94	6.94	4.23	22.15	7.46
21	3.08	0.89	7.06	4.12	18.62	4.45
22	3.24	0.83	8.12	5.07	15.98	5.41
23	2.85	0.81	6.64	4.03	19.10	6.13
24	3.64	0.84	8.02	4.61	14.41	4.72
25	3.06	0.80	7.85	4.47	18.57	3.97
26	3.22	0.98	7.29	4.82	23.17	6.70
27	3.37	0.79	7.51	4.90	20.70	7.66
28	3.33	0.82	7.81	4.72	14.21	5.64
29	3.06	0.86	7.13	4.30	17.18	4.34
30	3.11	0.83	7.72	4.93	12.89	4.42

Table 8.7. *WT simulaticn result (in minutes)*

According to Table 8.7 and Figure 8.25, the WT in both models of the vaccination center is relatively acceptable. The averages for the WT vary from 0.878 minutes in the proposed model with the theoretical arrivals scenario to 17.321 minutes in the real model with the real arrivals scenario. Furthermore, the difference between the WTs of the two models is significant, as indicated by the non-overlapping CIs of the WTs.

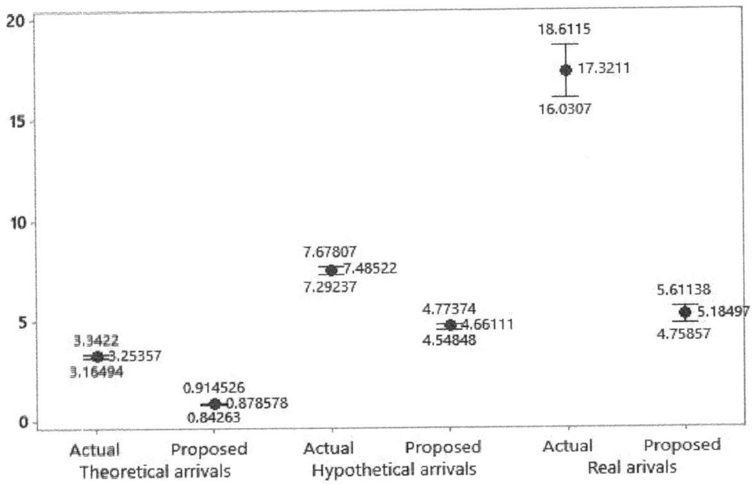

Figure 8.25. *The 95% CI for WT (in minutes)*

The Student's *t*-tests confirm that the WTs of the two center models differ significantly at the 0.05 significance level for the three scenarios (Figure 8.26). None of the CIs include the value of 0, and the p-values for all tests are less than 0.001.

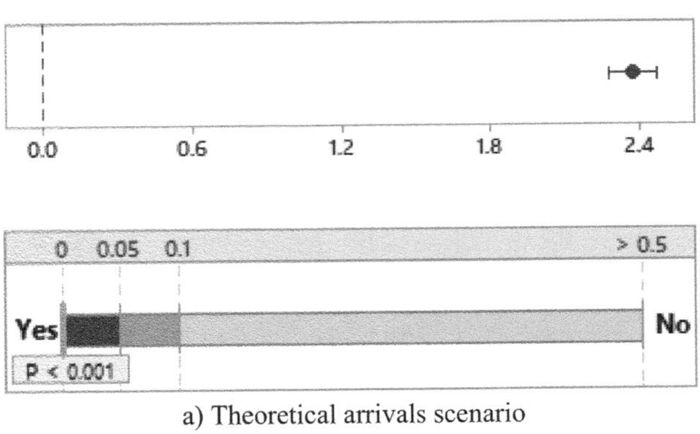

a) Theoretical arrivals scenario

Figure 8.26A. *Pairwise comparison of WT (in minutes) of the actual and proposed models. For a color version of the figure, see www.iste.co.uk/jerbi/discrete.zip*

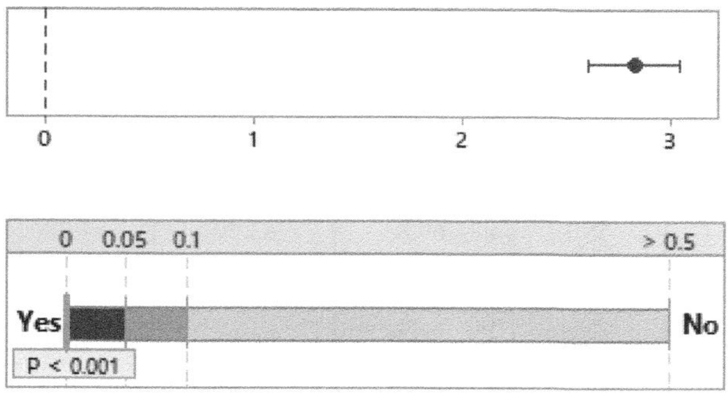

b) Hypothetical arrivals scenario

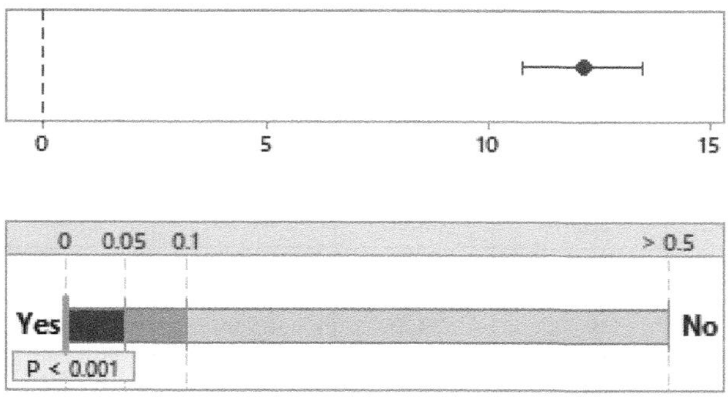

c) Actual arrivals scenario

Figure 8.26B. *Pairwise comparison of WT (in minutes) of the real and proposed models (continued). For a color version of the figure, see www.iste.co.uk/jerbi/discrete.zip*

8.10.2.4. *Comparison based on NCI*

Table 8.8 summarizes the simulation results of the average NCI in the two center models with the three arrival scenarios.

Rep	Theoretical arrivals		Hypothetical arrivals		Actual arrivals	
	Actual model	Proposed model	Actual model	Proposed model	Actual model	Proposed model
1	59.74	52.90	74.02	66.02	106.43	66.13
2	57.34	50.93	73.26	65.92	107.99	69.73
3	56.05	49.37	71.69	63.94	95.69	61.18
4	59.71	53.43	68.63	62.87	110.62	66.80
5	59.45	51.66	76.56	64.99	94.61	63.80

Rep	Theoretical arrivals		Hypothetical arrivals		Actual arrivals	
	Actual model	Proposed model	Actual model	Proposed model	Actual model	Proposed model
6	60.13	52.26	74.55	66.63	127.50	70.86
7	61.51	55.12	69.83	63.50	84.28	62.72
8	59.23	52.99	68.29	61.08	105.66	65.88
9	55.78	50.06	75.36	66.89	97.30	61.29
10	56.38	50.11	67.05	60.29	100.74	66.02
11	60.45	52.75	72.65	64.33	104.85	62.86
12	59.36	52.79	71.25	63.51	85.26	55.21
13	61.05	52.50	73.67	63.07	121.52	63.90
14	59.67	52.83	72.61	64.59	87.86	58.44
15	62.92	55.36	68.60	61.79	95.45	64.43
16	58.53	51.48	71.40	62.36	100.56	66.63
17	55.70	49.54	73.45	65.99	76.17	55.23
18	58.32	52.45	72.76	63.75	87.50	64.54
19	60.00	51.90	74.60	63.83	94.95	63.30
20	58.98	51.89	67.95	60.08	118.72	73.82
21	58.83	52.11	70.98	61.93	104.40	62.14
22	58.10	51.69	78.46	68.86	95.62	64.82
23	56.77	50.83	66.92	59.97	109.02	69.30
24	59.56	51.47	72.82	62.79	93.07	63.89
25	56.20	49.80	73.02	63.21	98.48	57.55
26	58.74	52.31	72.52	65.18	124.86	72.89
27	59.10	51.70	73.29	65.32	114.92	75.02
28	56.49	49.39	72.84	63.82	87.57	63.37
29	55.90	49.79	69.99	61.59	97.05	60.14
30	56.68	50.07	74.30	65.49	86.65	62.07

Table 8.8. *NCI simulation result (in citizens)*

Table 8.8 and Figure 8.27 show that both models present high numbers of citizens present in the center across all arrival scenarios. The average ranges from 51.716 to 100.510 citizens. However, the difference between the NCI of the two models is significant, as there is no overlap between the CIs for the NCI of the two models for any of the three arrival scenarios (Figure 8.27). The Student's *t*-tests confirm that the NCI of the two center models differs significantly at the 0.05 significance level for all three scenarios (Figure 8.28). None of the CIs include the value of 0, and the p-values of all tests are less than 0.001.

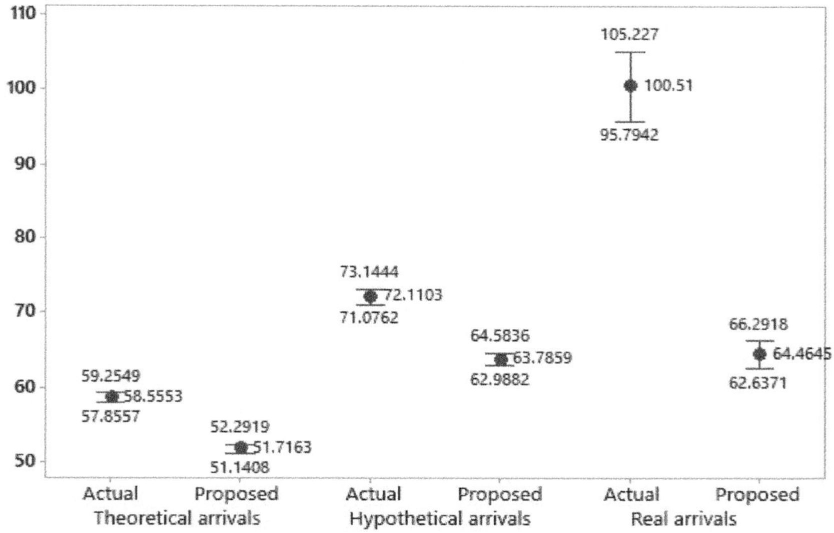

Figure 8.27. *The 95% CI for NCI (in citizens)*

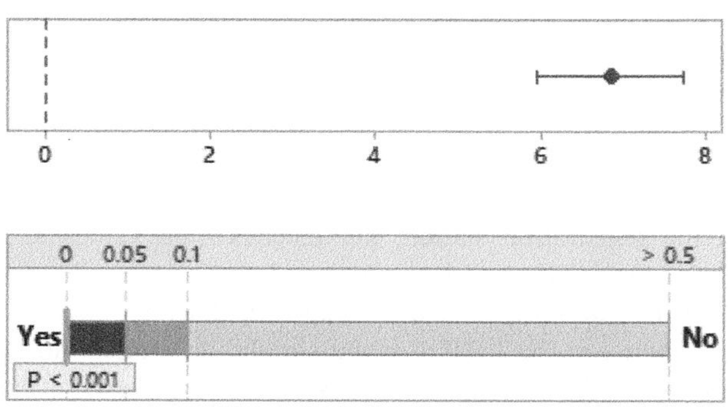

a) Theoretical arrivals scenario

Figure 8.28A. *Pairwise comparison of NCI (in citizens) of the real and proposed models. For a color version of the figure, see www.iste.co.uk/jerbi/discrete.zip*

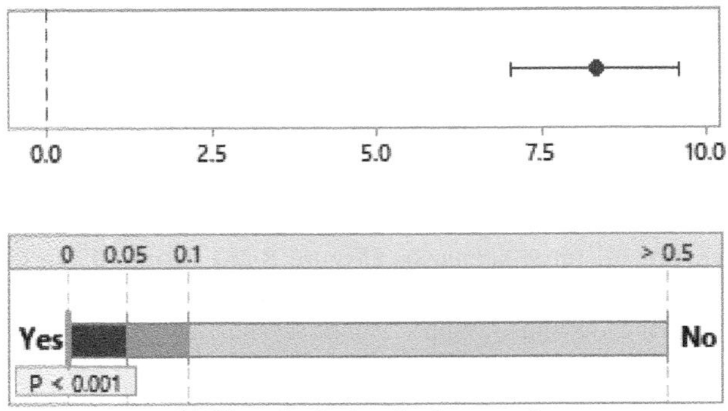

b) Hypothetical arrivals scenario

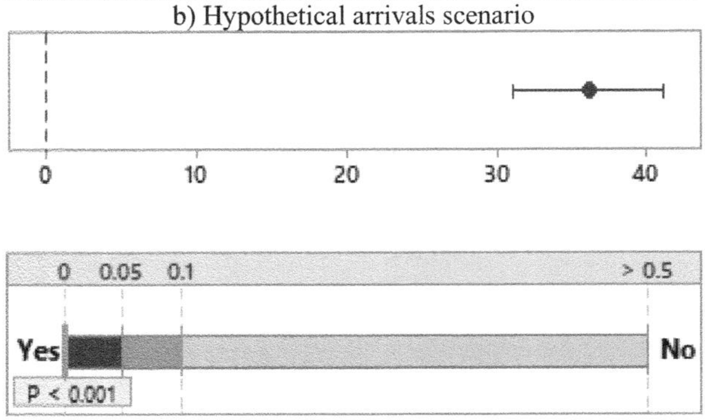

c) Actual arrivals scenario

Figure 8.28B. *Pairwise comparison of NCI (in citizens) of the real and proposed models (continued). For a color version of the figure, see www.iste.co.uk/jerbi/discrete.zip*

8.10.2.5. *Comparison based on UTIL*

The simulation results of the average operator utilization rate are provided in Appendix 4. Figure 8.29 presents the 95% CIs of this UTIL utilization rate per position for both models.

This average utilization rate varies across different positions. For instance, it remains low at the registration station, but reaches high values at the validation station. The CIs of the UTIL rate by model and scenario, as illustrated in Figure 8.30, indicate that the rate remains low as long as it does not exceed 50%. The CIs of the differences between the UTILs of the simulation models do not include the value of 0 for any of the three arrival scenarios. Moreover, *t*-tests reveal that the total UTILs of the actual and proposed models are significantly different at a 0.05 significance level for all three scenarios (Figure 8.31). The p-values of these tests are all less than 0.01, confirming the difference.

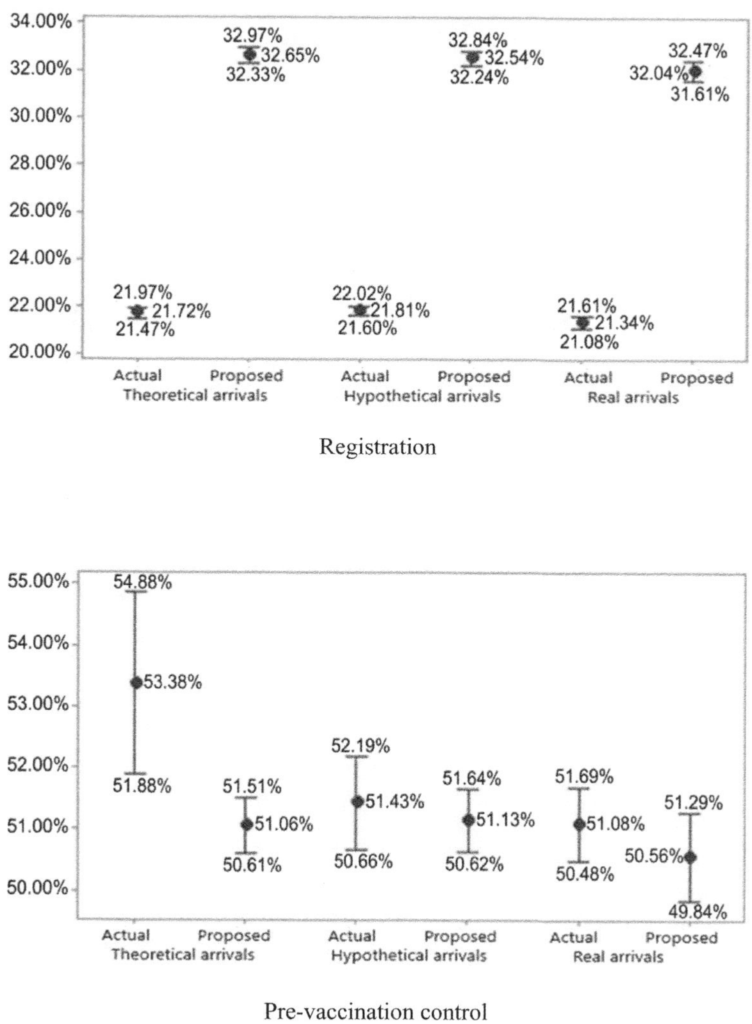

Figure 8.29A. *The 95% CIs of UTIL by position*

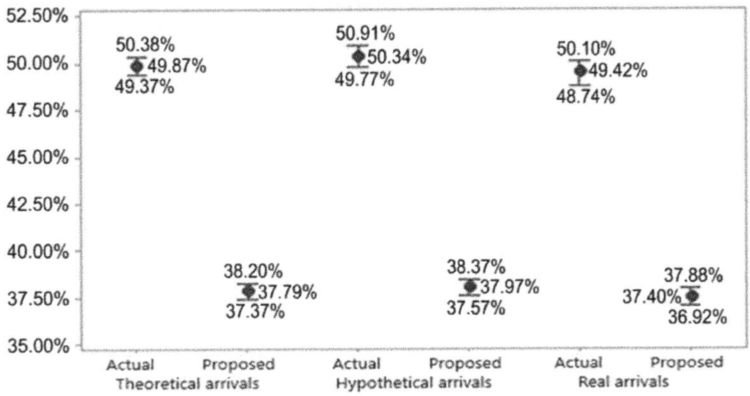

Vaccination

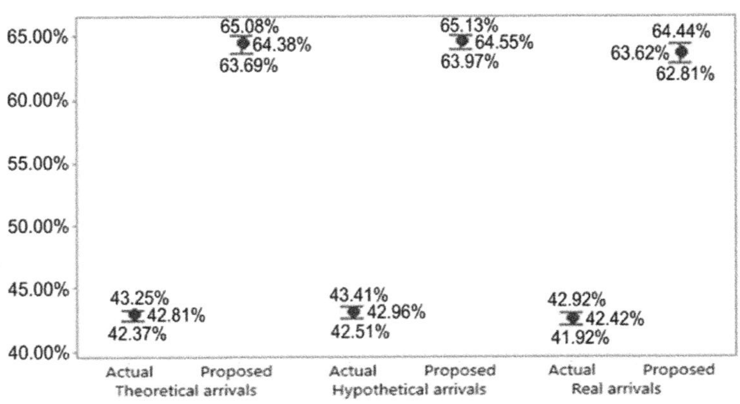

Validation

Figure 8.29B. *The 95% CIs of UTIL by position (continued)*

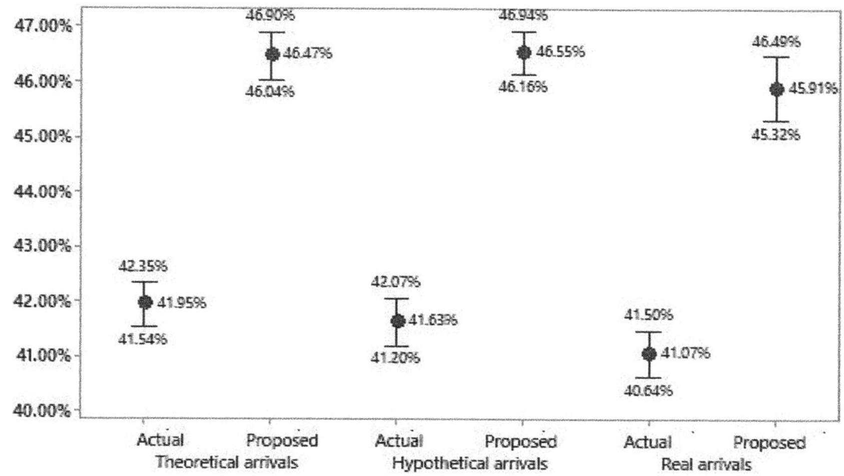

Figure 8.30. *The 95% CI for total UTIL (in %)*

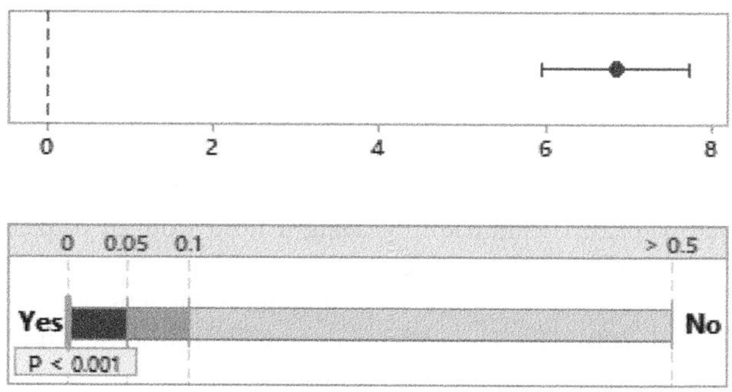

a) Theoretical arrivals scenario

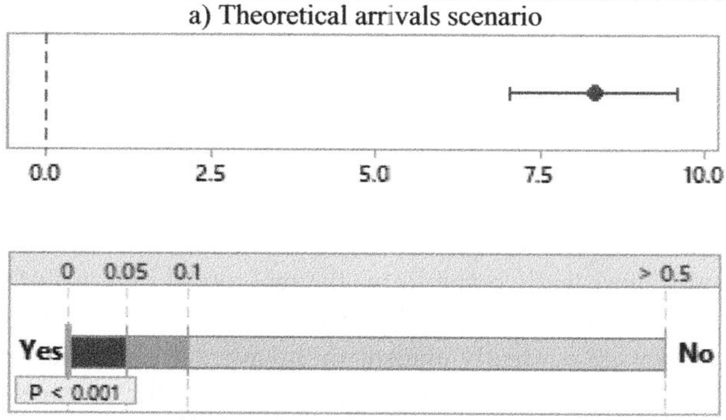

b) Hypothetical arrivals scenario

Figure 8.31A. *Pairwise comparison of UTIL (in %) of the real and proposed models. For a color version of the figure, see www.iste.co.uk/jerbi/discrete.zip*

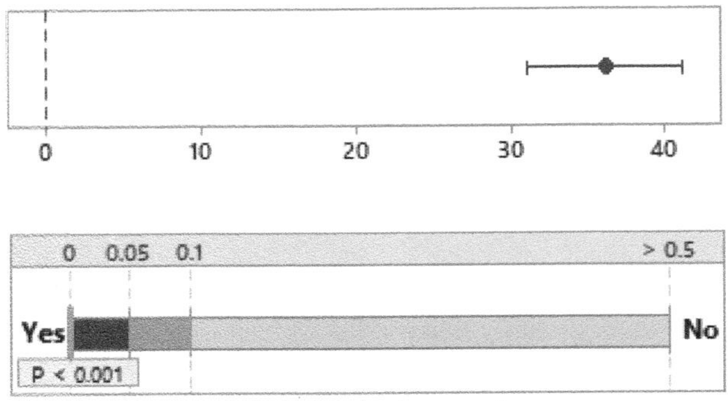

c) Actual arrivals scenario

Figure 8.31B. *Pairwise comparison of UTIL (in %) of the real and proposed models (continued). For a color version of the figure, see www.iste.co.uk/jerbi/discrete.zip*

8.11. Discussion and conclusion

8.11.1. *Center response to arrival variability*

The two simulation models of the vaccination center respond well to the variability of citizen arrivals. Specifically, 93.95% to 97.50% of citizens who arrive at the center receive the vaccine on time. In addition, the average total vaccination time does not exceed 33.839 minutes, which is a reasonable duration. The average waiting time remains at or below 17.321 minutes, indicating a fluid flow of citizens through the center. However, the main shortcoming in the center's management is the inability to maintain social distancing due to the large number of citizens present, which ranges between 51.716 and 100.510. This overcrowding compromises the adherence to social distancing guidelines.

8.11.2. *Comparison of the results of the two vaccination center models*

The percentages of citizens who receive the vaccine on time in the two models are not equivalent for the theoretical and hypothetical arrivals scenarios. However, the two simulation models behave equivalently with the real arrivals scenario. Notably, the proposed model consistently outperforms the real model in terms of the number of citizens vaccinated on time, even though the superiority is minor in the real arrivals scenario. This advantage is clearly evident in the average total vaccination time per citizen across all three arrival scenarios. In the real model, the average time increases significantly from one scenario to another, whereas in the proposed model, the increase is less pronounced. The difference in average vaccination times between the proposed and real model is 2.437, 2.813 and 12.141 minutes, respectively, for the theoretical, hypothetical and real arrivals scenarios, favoring the proposed model. In addition, the proposed model consistently generates a lower average waiting time than the real model across all three arrivals scenarios. The waiting time decreases by 73%, 37.72% and 70% in the proposed model compared to the real model for the theoretical, hypothetical and real arrivals scenarios, respectively.

The simulation results also show that the average number of citizens in the center increases from one scenario to another, with a more significant increase observed in the real model compared to the proposed model. In the real model, the average number of citizens rises from 58.555 for theoretical arrivals to 72.110 for hypothetical arrivals, a 25.28% increase. This number further escalates by 71.65%, reaching 100.51 citizens for real arrivals. In contrast, the proposed model exhibits a less pronounced increase in the average number of citizens. Specifically, the increase is 22.33% when transitioning from theoretical to hypothetical arrivals and 24.65% when moving from theoretical to real arrivals.

The average utilization rate of all operators in the real model is 41.95% for theoretical arrivals, 41.63% for hypothetical arrivals and 41.077% for real arrivals. In contrast, the proposed model shows higher utilization rates of 46.47%, 46.55% and 46.49% for the respective scenarios. This indicates that operators are utilized more effectively in the proposed model compared to the real model.

8.11.3. *Conclusion*

Both simulation models demonstrate robust responsiveness to the variability in citizen arrivals, showcasing the effectiveness of their operational frameworks. However, a comprehensive analysis of performance indicators reveals that the proposed model, which includes reorganized tasks and optimized operator assignments, consistently outperforms the existing vaccination center model.

The proposed model's strategic restructuring of tasks and streamlined operator allocation contribute significantly to its enhanced performance metrics across various scenarios. Key indicators such as the percentage of citizens vaccinated on time, average total vaccination time per citizen and average waiting time are notably superior in the proposed model compared to the real-world center.

Despite these advancements, an area for improvement in the proposed model lies in optimizing the average utilization rates of operators. While they are slightly higher than those in the real model, further enhancements in operator efficiency could yield even better outcomes, potentially reducing operational costs and improving overall service delivery.

In conclusion, while the proposed model demonstrates clear advantages over the real-world vaccination center in terms of efficiency and performance metrics, ongoing refinements are necessary to maximize operator utilization and achieve optimal operational excellence.

9

Simulation in Supply Chains: Optimizing CO_2 Emissions Through Pooling Strategies

9.1. Introduction

The rapid pace of mass industrialization and the trend towards excessive consumption are the primary drivers of the significant climate change we are experiencing. This drastic shift is leading to increasingly frequent natural disasters that disrupt supply chains. Greenhouse gas (GHG) emissions from human activities are the main cause of this climate change. As populations, economies and living standards rise, the cumulative level of GHGs also grows. Moreover, recent studies indicate that many consumers now prefer options that reduce carbon emissions. The most prevalent GHG is carbon dioxide (CO_2), predominantly produced by the combustion of fossil fuels, particularly in logistics activities. In this context, collaboration between supply chains has become essential. Such collaboration enhances managers' visibility and helps them to decide on the best transport strategies to minimize CO_2 emissions from their supply chains.

The pooling strategy within supply chains stand out as an innovative approach to enhancing collaboration. This strategy transcends the conventional practice of simply consolidating vehicles, which is quite common in the industry. It is fundamentally rooted in a partnership agreement where multiple supply chains voluntarily pool and share resources (infrastructure, vehicles, etc.) and information to achieve specific performance objectives or benefits. Initially, the pooling of logistics resources was driven solely by economic incentives. However, with the advent of stricter environmental regulations, logisticians have increasingly adopted this strategy for its ecological benefits, including the significant reduction of CO_2 emissions.

9.2. Pooling strategy

Transport sharing encompasses a range of definitions, but the common thread is that it is a collaborative agreement among multiple stakeholders. This arrangement involves the voluntary sharing of physical resources, information and expertise to achieve collective economic, ecological and financial benefits that are impossible to achieve independently. The structure of this cooperation can take various organizational forms, depending on the nature of the involved parties, the resources and the products or services in question.

9.2.1. *Multi-pick pooling strategy (pick-up round)*

This strategy involves consolidation of deliveries bound for the same destination (customer) from multiple shipping points (suppliers). As a result, orders from several suppliers intended for the same customer are combined through a collection tour performed by one or more vehicles. Figure 9.1 illustrates a conceptual supply chain with multiple pick-up points.

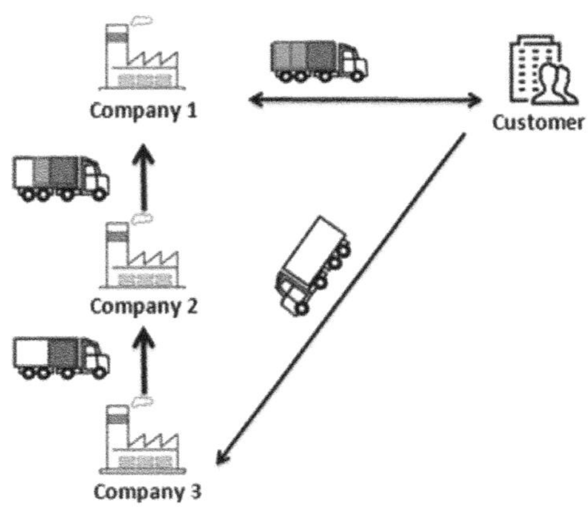

Figure 9.1. *Structure of a pooled supply chain with the multi-pick strategy. For a color version of the figure, see www.iste.co.uk/jerbi/discrete.zip*

9.2.2. *Multi-drop pooling strategy (distribution tour)*

The multi-drop strategy follows a similar principle to multi-pick, but it begins with consolidating orders from a single origin and distributing from multiple destinations. Several vehicles depart from a common supplier to transport various orders, delivering them sequentially to several customers along a distribution tour. These customers are typically in close proximity geographically or along the same delivery route. Figure 9.2 illustrates a conceptual supply chain employing the multi-drop strategy.

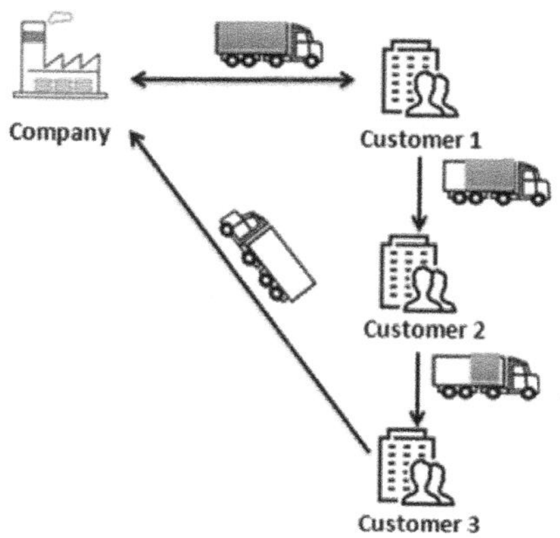

Figure 9.2. *Structure of a pooled supply chain with the multi-drop strategy. For a color version of the figure, see www.iste.co.uk/jerbi/discrete.zip*

9.2.3. *Pooling strategy with pooled hub*

The supply chain incorporates one or more pooled intermediate hubs, facilitating deliveries in two stages. Suppliers direct their products to these hubs, and when customers specify their delivery needs, orders are dispatched from the hub for final delivery.

Integrating these hubs into the supply chain structure enhances shipment sizes and optimizes transportation costs by maximizing truck fill rates. Figure 9.3 depicts a conceptual supply chain with shared a pooled hub.

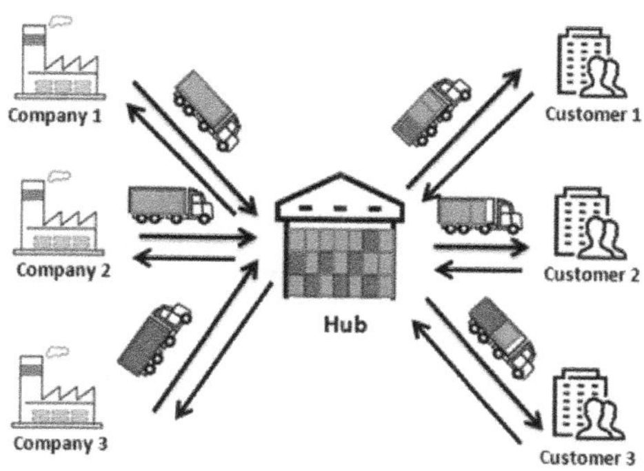

Figure 9.3. *Structure of a pooled supply chain with pooled hub. For a color version of the figure, see www.iste.co.uk/jerbi/discrete.zip*

9.3. Objective

This simulation project aims to evaluate how different pooling strategies within the supply chain can effectively reduce CO_2 emissions using discrete event simulation.

9.4. Description of the channel studied

The simulated supply chain involves two companies named "company i" and three customers named "customer i". These companies supply products to the supply chain based on statistical distributions detailed in Table 9.1.

		Inter-arrival time (hour)	Quantity of pallets
Company 1	Product 1	N(4;0.5)	2
	Product 2	N(3;0.5)	2
Company 2	Product 3	N(3.5;0.5)	2
	Product 4	N(3.5;0.5)	2

N: normal distribution.

Table 9.1. *Pallet supply time and quantity*

Each customer places separate delivery orders with both companies on varying dates. Inter-order times and the corresponding pallet quantities for each product type are governed by discrete distributions (Table 9.2).

	Inter-order time	Quantity of pallets per order		
Customer 1	U[12; 24]	Company 1	Product 1	DISC (0.6; 4; 1.0; 5)
			Product 2	DISC (0.3; 5; 1.0; 7)
		Company 2	Product 3	DISC (0.8; 6; 1.0; 7)
			Product 4	DISC (0.7; 4; 1.0; 6)
Customer 2	U[12; 24]	Company 1	Product 1	DISC (0.4; 4; 1.0; 5)
			Product 2	DISC (0.6; 5; 1.0; 7)
		Company 2	Product 3	DISC (0.5; 3; 1.0; 5)
			Product 4	DISC (0.8; 4; 1.0; 6)
Customer 3	U[24; 36]	Company 1	Product 1	DISC (0.5; 3; 1.0; 6)
			Product 2	DISC (0.5; 4; 1.0; 8)
		Company 2	Product 3	DISC (0.3; 3; 1.0; 6)
			Product 4	DISC (0.3; 6; 1.0; 8)

U: uniform distribution; DISC: discrete distribution.

Table 9.2. *Time between orders and quantity of pallets per order*

Distances between components within the supply chain are specified in kilometers in Table 9.3. Average delivery vehicle speeds are determined by the distance between origin and destination. For longer distances, the average speed is set at 80 km per hour, while for shorter and medium distances, it is reduced to 40 km per hour (Table 9.3). Vehicle loading adheres to recommendations from the Environment and Energy Management Agency, limiting maximum capacity to 6 tons due to pallet volume constraints. Each company operates a fleet of two trucks. In models involving a pooled hub, these trucks are allocated between the companies and the hub. Consequently, companies use a shared truck to transport pallets to the consolidation center while the remaining three trucks serve the hub for customer deliveries.

Hub	*					
Company 1	D: 35, v: 40	*				
Company 2	D: 46, v: 40	D: 11, v: 40	*			
Customer 1	D: 155, v: 80	D: 190, v: 80	D: 201, v: 80	*		
Customer 2	D: 180, v: 80	D: 210, v: 80	D: 221, v: 80	D: 20, v: 40	*	
Customer 3	D: 245, v: 80	D: 280, v: 80	D: 291, v: 80	D: 90, v: 80	D: 55, v: 40	*
	Hub	Company 1	Company 2	Customer 1	Customer 2	Customer 3
	D: distance in kilometers, v: speed in kilometers per hour					

Table 9.3. *Distances and speeds in the supply chain*

9.5. Formulation of the CO_2 emissions indicator

The calculation of CO_2 emissions is influenced by factors such as the transported load weight, vehicle capacity and travel distance. The equations for estimating CO_2 emissions during delivery are detailed as follows:

$$E_{loaded} = D_{ij} \times [E_{empty} + (E_{full} - E_{empty}) \times X] \quad [9.1]$$

$$E_{unloaded} = D_{ij} \times E_{empty} \quad [9.2]$$

where:

– E_{loaded}: the total CO_2 emissions of a loaded vehicle ($kgCO_2$) between nodes i and j;

– $E_{unloaded}$: the total CO_2 emissions of an unloaded vehicle ($kgCO_2$) between nodes i and j;

– D_{ij}: the distance in kilometers between nodes i and j;

– E_{empty}: the CO_2 emission rate ($kgCO_2$/km) of an empty vehicle;

– E_{full}: the CO_2 emission rate ($kgCO_2/km$) of a fully loaded vehicle;

– X: the ratio between the weight of the cargo load and the maximum payload of the vehicle.

CO_2 emission rates of empty and fully loaded vehicles E_{empty} and E_{full} are linked to the speed of the vehicle in the following way:

$$E_{empty} = K + a \times v + b \times v^2 + c \times v^3 + d/v + e/v^2 + f/v^3 \quad [9.3]$$

$$E_{full} = E_{empty} \times C_{loaded}(v) \quad [9.4]$$

$$C_{loaded}(v) = k + r \times v + s \times v^2 + t \times v^3 + u/v \quad [9.5]$$

where:

– V: vehicle speed;

– K and k: constants determined according to the gross vehicle weight rating (GVWR) of the vehicle;

– a, b, c, d, e, f, r, s, t, u: coefficients determined according to the vehicle's GVWR;

– C_{loaded}: the load correction factor.

Using the truck's GVWR and the previous equations, the values of the full and empty emission rates are estimated. The calculated values are E_{empty} = 0.48 $kgCO_2/km$ and E_{full} = 0.58 $kgCO_2/km$ at a speed of 40 km/h and E_{empty} = 0.51 $kgCO_2/km$ and E_{full} = 0.58 $kgCO_2/km$ at a speed of 80 km/h. Each truck has an average load capacity of 9.8 tons.

9.6. Conceptual models

9.6.1. *Non-pooled supply chain*

In this model, customers initiate purchase orders, sending them to the company with a copy to their receiving department. The company manufactures and stores products until customer orders are received. Upon order receipt, stock availability is verified. If sufficient stock is available, the order is prepared and directed to the company's shipping dock. A suitable truck, meeting the required load capacity, is assigned to transport the order to the customer's unloading dock for receipt. If the order exceeds the initial truck's capacity, additional trucks may be deployed as needed. Figure 9.4 illustrates the various process flows within this chain model, with physical flows represented by continuous lines and informational flows represented by dashed lines.

9.6.2. *Pooled supply chain with multi-pick strategy*

The particularity of this supply chain is that a customer's order may be fulfilled by multiple companies, requiring coordinated efforts between them. This coordination is managed by the "multi-pick management" section within each company, which assigns and schedules the appropriate truck(s) and coordinates their routes (Figure 9.5).

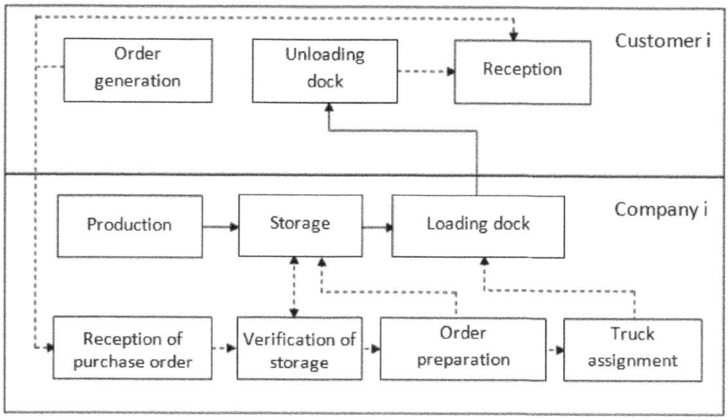

Figure 9.4. *Conceptual model of the non-pooled supply chain*

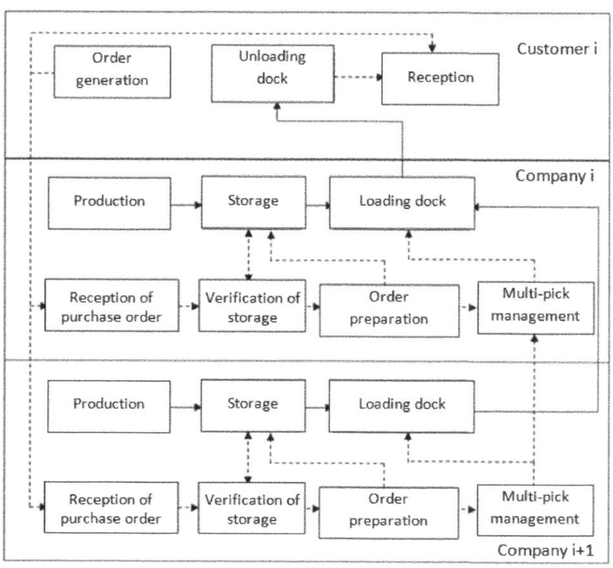

Figure 9.5. *Conceptual model of the pooled supply chain with multi-pick strategy*

9.6.3. *Pooled supply chain with multi-drop strategy*

This supply chain's distinctive aspect is its ability to consolidate orders from multiple customers for delivery by a single company. Effective coordination among these customers is essential. This coordination is managed by the respective "multi-drop management" sections within each customer and company, overseeing and coordinating deliveries among customers (Figure 9.6).

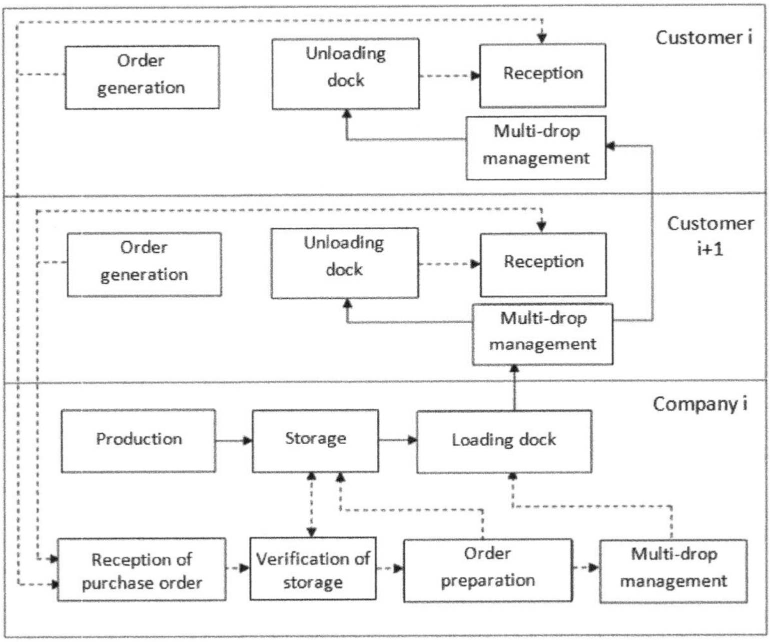

Figure 9.6. *Conceptual model of the pooled supply chain with multi-drop strategy*

9.6.4. *Pooled supply chain with hub*

This supply chain operates across three tiers: the company, the hub and the customer. Customer orders are no longer sent to the company but are routed to a central pooling hub. This hub consolidates various products through regular deliveries from companies and manages the fulfillment of orders to customers. Consequently, functions such as "Receipt of purchase order", "Stock check", "Order preparation" and "Truck assignment" are relocated from the companies to the hub (Figure 9.7).

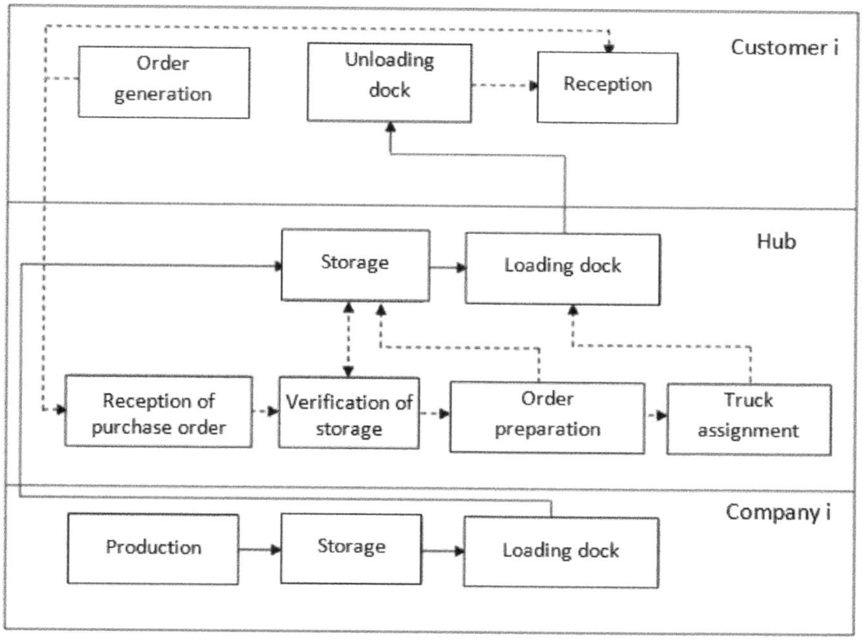

Figure 9.7. *Conceptual model of the pooled supply chain with hub*

9.7. Simulation models

All simulation models were developed using the Arena 14.0 simulation software.

9.7.1. *Non-pooled supply chain model*

The structure of the non-pooled supply chain model consists of two main components: companies and customers. The company components are divided into two sub-models, each representing an individual company. The customer component comprises three sub-models, each representing a specific customer (Figure 9.8).

Each company sub-model includes three key sections "Company i: order preparation", "Company i: customers assignment" and "Company i: loading docks" (Figure 9.9).

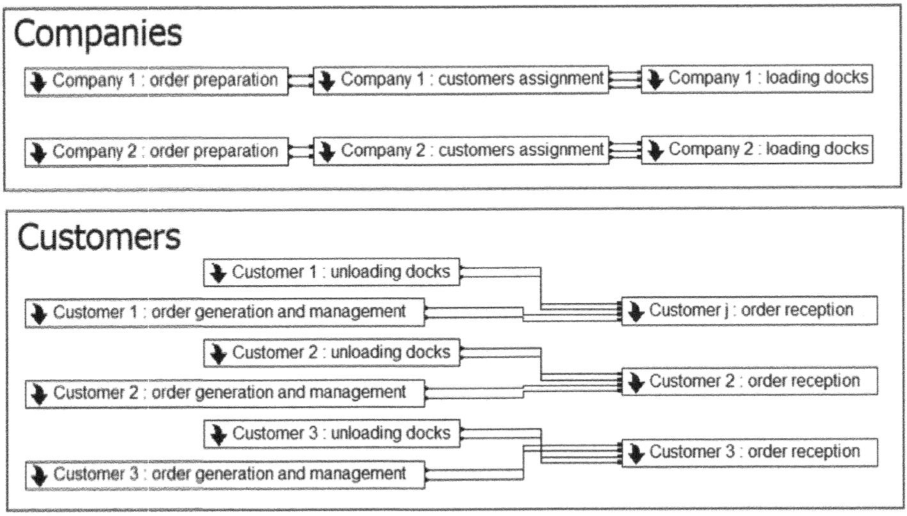

Figure 9.8. *Non-pooled supply chain simulation model*

The first section acts as a generator of product pallets and manages order preparation. Each company produces two types of products, with dedicated "Create" modules for each product. Each product pallet is held in a storage area, represented by a "Hold" module until a customer order is received. When an order is placed, the corresponding customer sends a signal to the appropriate "Create" module to release the required number of pallets. These pallets then pass through a "Decide" module to verify if their quantity meets the vehicle loading capacity. If the quantity is equal or greater than the vehicle, the pallets corresponding to this capacity are grouped into a full load by the "full load preparation company i" "Batch" module. This load proceeds to an "Assign" module, where it receives attributes and variables related to load size and order size. The order size variable is then reduced by the number of pallets already grouped. If the load condition is not met, all pallets are sent to the "Partial load preparation company i" "Batch" module. Before this, a variable indicating the remaining part of the order is incremented as the pallets through the "Remaining load company i reset" "Assign" module. The load then goes through additional "Assign" modules to update the "load size" attribute and reset all other variables.

In the second section, "Company i: Customer Assignment", each delivery is assigned to the corresponding customer through a "Decide" module, which is connected to three "Assign" modules, each corresponding to a customer. These modules assign the necessary attributes, such as order number, number of pallets and delivery order reason, to each delivery before loading it onto a vehicle. The delivery sequences are pre-defined in the “Sequence” modules.

In the third section, “Company i: Loading docks”, each delivery waits for the nearest available vehicle in the queue of a “Request” module. The vehicle picks up the load at the “Company i” station and proceeds to the “Transport” module, where the travel speed is defined before being sent to the respective customer.

Similar to the company sub-model, each sub-model of a customer includes three sections: “Customer j: order generation and management”, “Customer j: unloading dock” and “Customer j: order reception” (Figure 9.10).

In the “Customer j: order generation and management” section, an initial “Create” module generates orders, which are then duplicated based on the number of companies by a “Separate” module named “Duplication order customer j”. Each duplicated order is processed by an “Assign” module. These modules increment the order number and assign the required number of pallets per product type to each order. Subsequently, each order passes through a “Decide” module named “Storage Company i Customer j” to check the availability of the pallets in the companies’ storage. If pallets are available, the order is validated and a signal is sent via a “Signal” module to the corresponding company to release the requested number of pallets. Conversely, if pallet availability is insufficient, the order is canceled and directed to a “Dispose” module.

In the “Customer j: loading docks” section, each vehicle entering is directed to a “Free” module. This module releases the vehicles from their loads, which are transmitted to the “Record” modules. The first records the total number of delivered pallets, while the second calculates CO_2 emissions.

In the “Customer j: order reception” section, each delivered load is separated into individual pallets by a “Separate” module. These pallets are subsequently routed to a “Batch” module, which groups all pallets belonging to each order. Once all ordered pallets are grouped together, a “Match” module matches them with the corresponding order. To confirm order receipt, the paired informational (order) and physical (pallet) flows are grouped by a “Batch” module and subsequently removed from the system via a “Dispose” module.

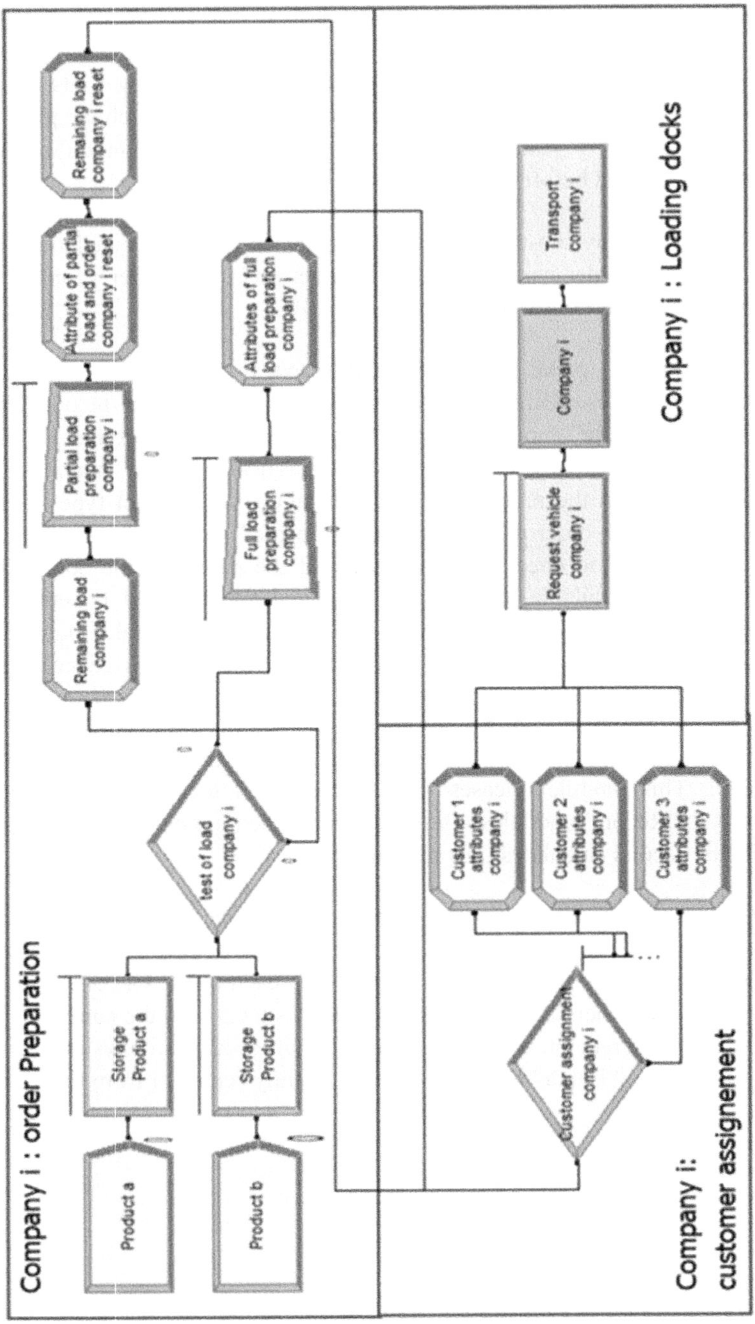

Figure 9.9. *Simulation model of a "Company i" section for the non-pooled supply chain. For a color version of the figure, see www.iste.co.uk/jerbi/discrete.zip*

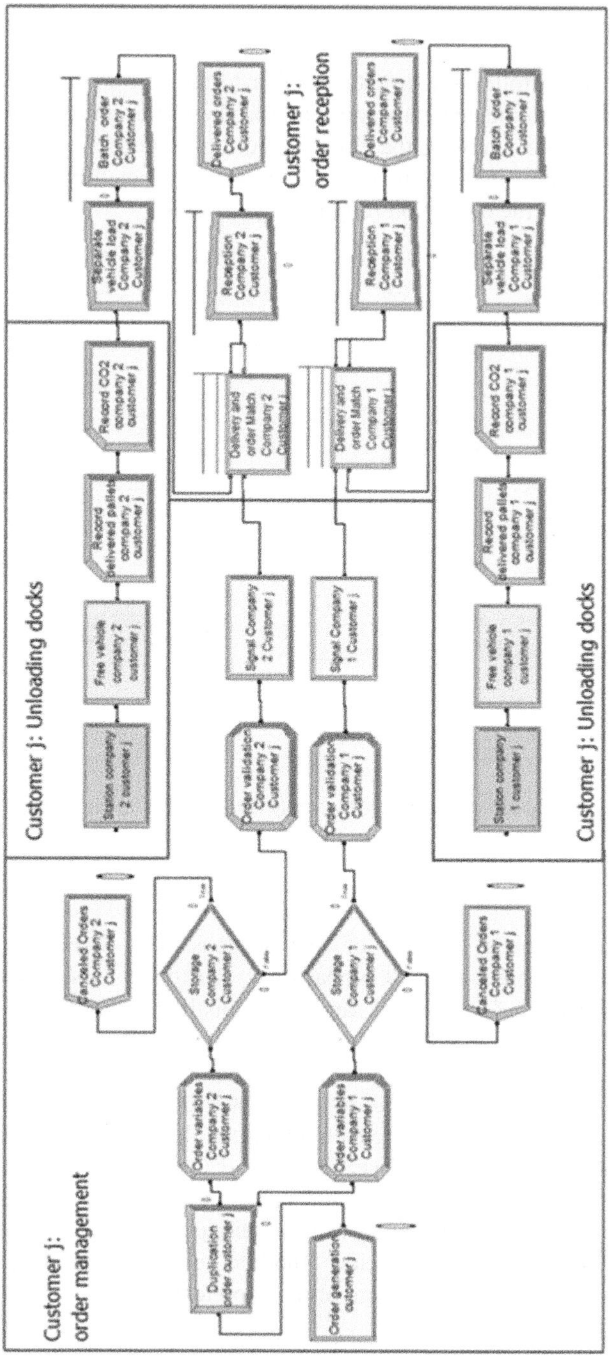

Figure 9.10. *Simulation model of a "Customer j" section for the non-pooled supply chain. For a color version of the figure, see www.iste.co.uk/jerbi/discrete.zip*

9.7.2. *Pooled supply chain model with multi-pick strategy*

This simulation model follows a structure similar to the non-pooled model, with the main difference lying in the company sub-models. These sub-models now include a new section dedicated to handling the flow of multi-pick deliveries. In addition, each company sub-model is structured differently to efficiently manage the reception of pallets during the collection rounds (Figure 9.11).

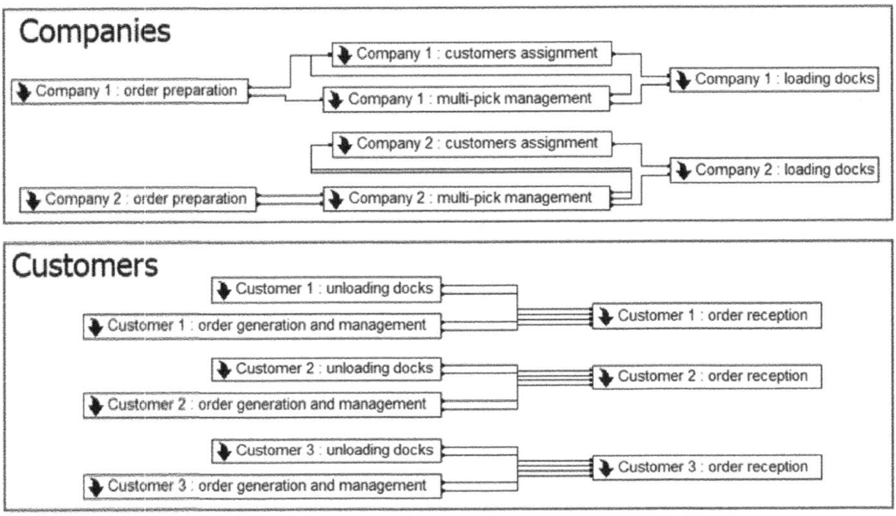

Figure 9.11. *Simulation model of the pooled supply chain with the multi-pick strategy*

9.7.2.1. *Company 1 sub-model*

The company 1 sub-model consists of four sections. The first three sections "Company 1: order preparation", "Company 1: customer assignment" and "Company 1: loading docks" retain the same functionality and structure as their counterparts in the company sub-model of the non-pooled supply chain model. The fourth section, titled "Company 1: multi-pick management", specifically handles the coordination of multi-pick deliveries (Figure 9.12). After delivering a vehicle with a full capacity load to customers, any remaining partial loads are directed to the "Company 1: multi-pick management" section. These loads are held in a "Hold" module, named "Company 1 partial load queue", awaiting a signal from company 2 or the same customer, indicating the application of the multi-pick strategy. Upon receiving, the partial load passes through a "Decide" module to determine its path to a sub-section based on the feasibility of the multi-pick application.

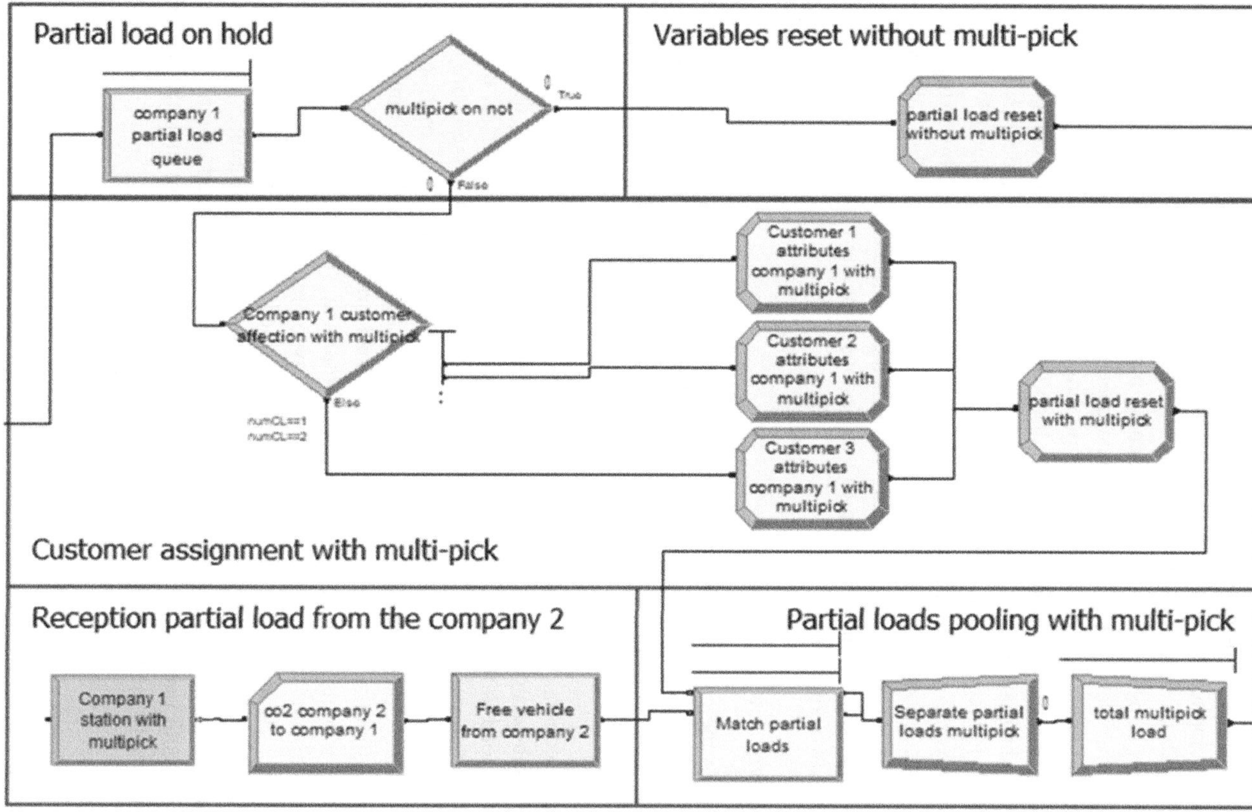

Figure 9.12. *Simulation model for "Company 1: multi-pick management" sections. For a color version of the figure, see www.iste.co.uk/jerbi/discrete.zip*

If the condition for applicability is not met, the multi-pick process is canceled and the remaining load passes through an “Assign” module in the “Variable reset without multi-pick” sub-section to reset all corresponding variables. Conversely, if multi-pick is applicable, the load is directed to the “Customer assignment with multi-pick” sub-section. Here, the load is assigned to the respective customer along with necessary transfer attributes. Subsequently, it moves to another “Assign” module to reset all other variables. The load is then queued in the “Match” module, named “Match partial loads”, in the “Partial loads pooling with multi-pick” sub-section. The primary function of this module is to match partial loads from both companies for the corresponding customer. These loads are disaggregated into individual pallets and subsequently aggregated into a single load by the subsequent “Batch” module. Simultaneously, the second queue of the “Match” module is populated by partial loads from the “Reception partial loads from Company 2” sub-section. This sub-section receives partial loads sent by company 2 when the multi-pick strategy applies. In addition, it calculates and stores the CO_2 emissions generated during load transfers between the two companies.

9.7.2.2. *Company 2 sub-model*

Similar to the company 1 sub-model, the company 2 sub-model comprises four sections. Each section mirrors the structure and functionality of the corresponding sections in the company 1 sub-model, except for the “multi-pick management” section (Figure 9.13).

In this section, each complete load enters the “Check of the multi-pick feasibility” sub-section, where the “Decide” module, named “Last load”, determines whether this load is the last one scheduled for the order’s delivery. If it is, the load moves to the “no multi-pick” sub-section to reset variables related to the entire order and then passes through a “Signal” module to notify Company 1 that multi-pick is not applicable. Otherwise, the entire load is sent directly to the “Company 2: customers assignment” section. When a partial load arrives in the “Check of the multi-pick feasibility” sub-section, it passes through a “Decide” module, named “Multi-pick or not”, which assesses the feasibility of multi-pick and searches for another partial load waiting in Company 1 destined to the same customer. A subsequent check, named “Vehicle and multi-pick load compatibility”, ensures compatibility between the vehicle’s load capacity and the combined quantity of the pallets from both partial loads. If the compatibility condition is not met, the partial load is directed to the “Multi-pick number” sub-section.

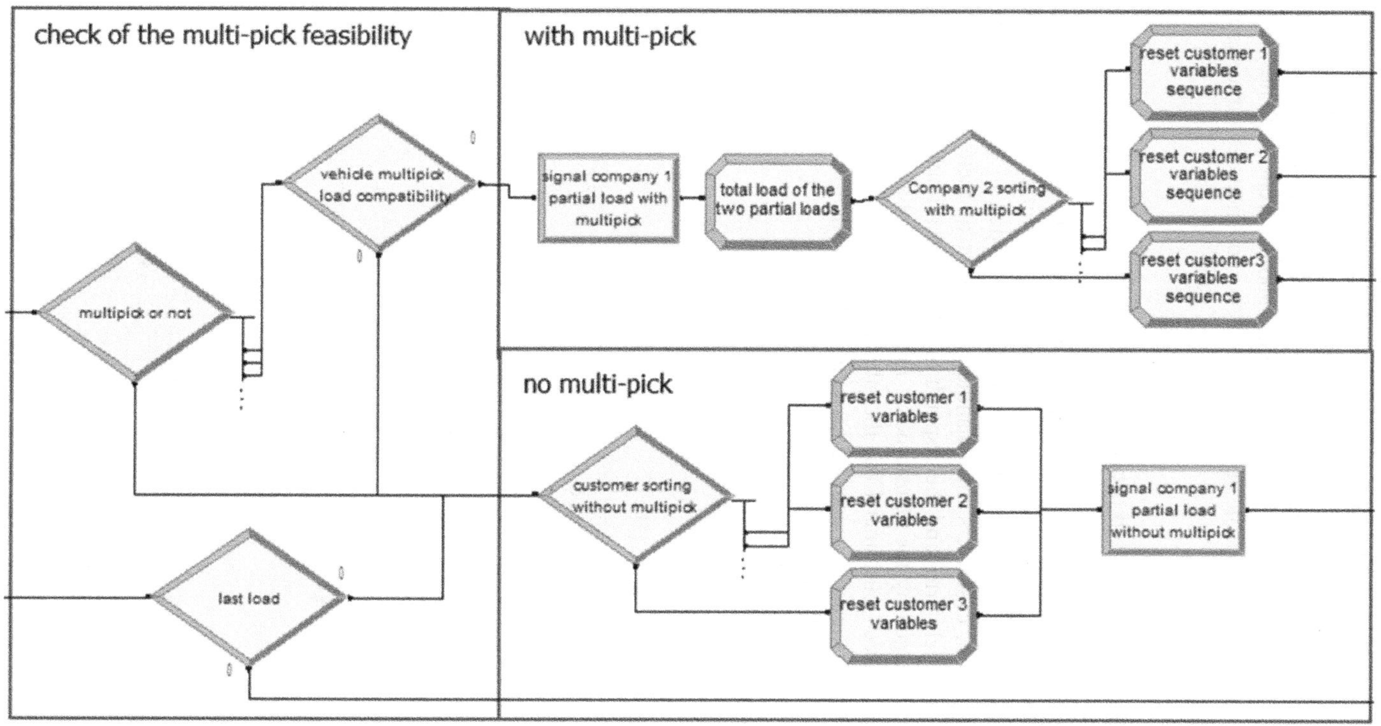

Figure 9.13. *Simulation model of the "Company 2: multi-pick management" sections*

Otherwise, compatibility is validated and the partial load proceeds to the "Signal" module, which instructs company 1 to release the partial load waiting in the "Company 1 multi-pick management" section. The load then moves through the "Assign" module to update different variables and different attributes. Finally, it is directed to a "Decide" module and a series of "Assign" modules for assignment to the appropriate customer.

9.7.3. *Pooled supply chain model with multi-drop strategy*

In this simulation model, each company is responsible for managing the joint delivery of multiple parts to various customers. To facilitate this, a new section named "Company i: multi-drop management" is included in each company sub-model. In addition, customers must handle the receipt of their partial loads from these multi-drop deliveries and the dispatch of other partial loads to their customers. This task is managed by the "Customer j: multi-drop management" section. These sections vary between Customer 1 and the other customers, as Customer 1 is closest to the companies and must manage the reception and dispatch of a larger number of partial loads (Figure 9.14).

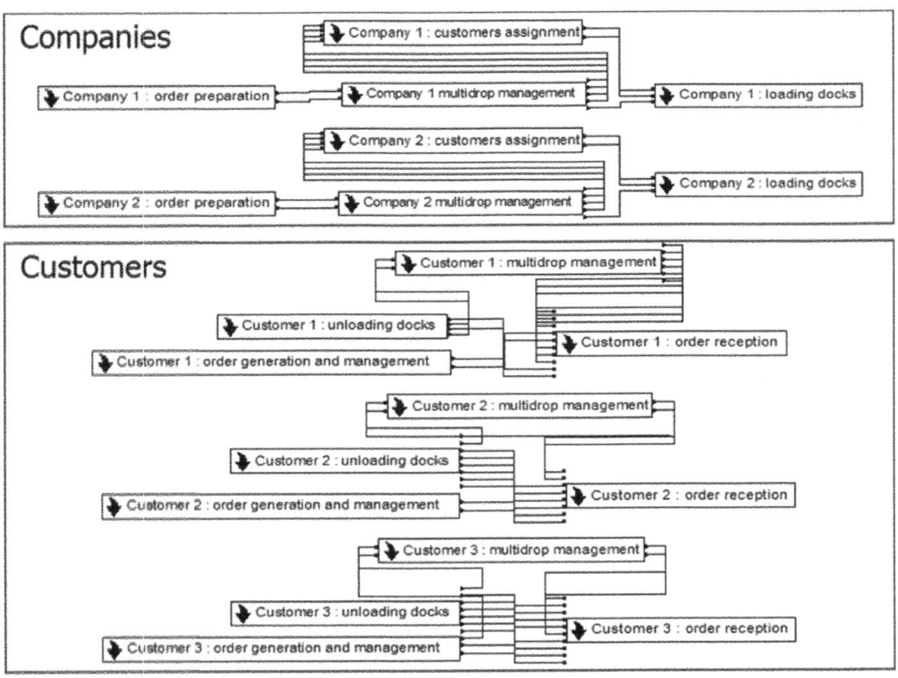

Figure 9.14. *Simulation model of the pooled supply chain with the multi-drop strategy*

9.7.3.1. *Company sub-model*

Each sub-model includes four sections. The sections "Company j: order preparation", "Company j: customers assignment" and "Company j: loading docks" maintain the same structures as their counterpart sections in the simulation model without pooling. In the "Company j: multi-drop management" section, all loads are kept in the queue of the "Hold" module named "Load queue". This module holds the loads until the pending load in the "Request truck company j" module of the "Company j: loading docks" section is assigned to a free vehicle. Once released from the queue, the load goes through two successive loops named "Partial load search" to consolidate the waiting partial loads and to optimize the vehicle filling rate for any multi-drop delivery (Figure 9.15).

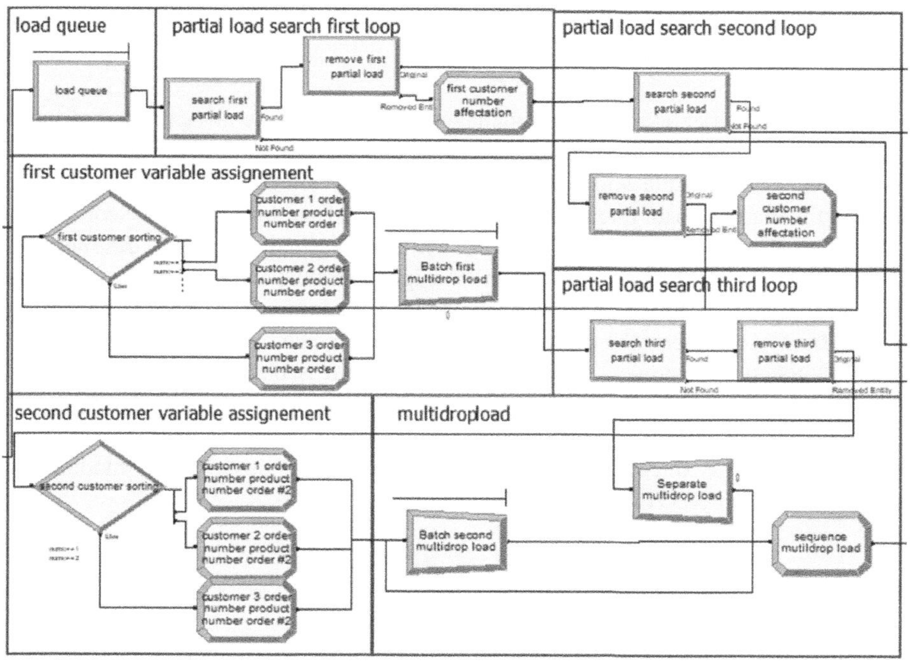

Figure 9.15. *Simulation model of the "Company j: multi-drop management" section*

In the first loop, a "Search" module identifies a pending partial load. Once found, a "Remove" module named "Remove first partial load" takes this load from the "Hold" queue. The original load is then directed to the "Company j: customer assignment" section. The removed partial load is directed to an "Assign" module, where variables related to the customer and the load are updated. This first partial load then goes through a second "Search" module, in the second loop, to find another partial load in the "Hold" queue. The combined number of pallets of these two partial loads must not exceed the vehicle's capacity. If no suitable load is found, the first partial load is sent to the "Company j: customer assignment" section. Otherwise, the second partial load is removed from the "Hold" queue by the "Remove" module named "Remove the second partial load" and transferred to the "Assign" module to update the customer variables and load again. The two partial loads are then sent to the "First customer assignment" sub-section for assignment to the corresponding customer and grouped. Another cycle will search for a third partial load to further optimize the vehicle's capacity. All grouped partial loads are then sent to the "Company j: load docks" section.

9.7.3.2. *Customer sub-model*

In the "Customer 1: multi-drop management" section, the grouped partial loads are sorted by a "Decide" module according to their destinations. They are separated by a "Separate" module and further sorted again by another "Decide" module to extract the partial load intended for Customer 1. This load is then sent to the "Customer 1: order reception" section, while the remaining load, destined for another customer, proceeds through an "Assign " module where it is assigned the necessary attributes for the next delivery. If the remaining load needs to be sent to two different customers, the multi-drop load is recalculated in the "Assign" module named "Direction customer 3 customer 2". This load is batched again and sent to another Assign module, where the new transfer sequence and other attributes are assigned and the multi-drop variables are reset in a subsequent "Assign" module. Finally, the load is directed to the "Station multi-drop to customer 3" module for delivery to customer 3 (Figure 9.16).

9.7.4. *Pooled supply chain model with hub*

Unlike previous models, this model consists of three sub-models: the production companies sub-model, which includes two sections, each representing a company; the customers sub-model, which includes three sections, each corresponding to a customer; and the hub model. Intermediate deliveries are routed from the companies to the hub, which is responsible for supplying all customers (Figure 9.17).

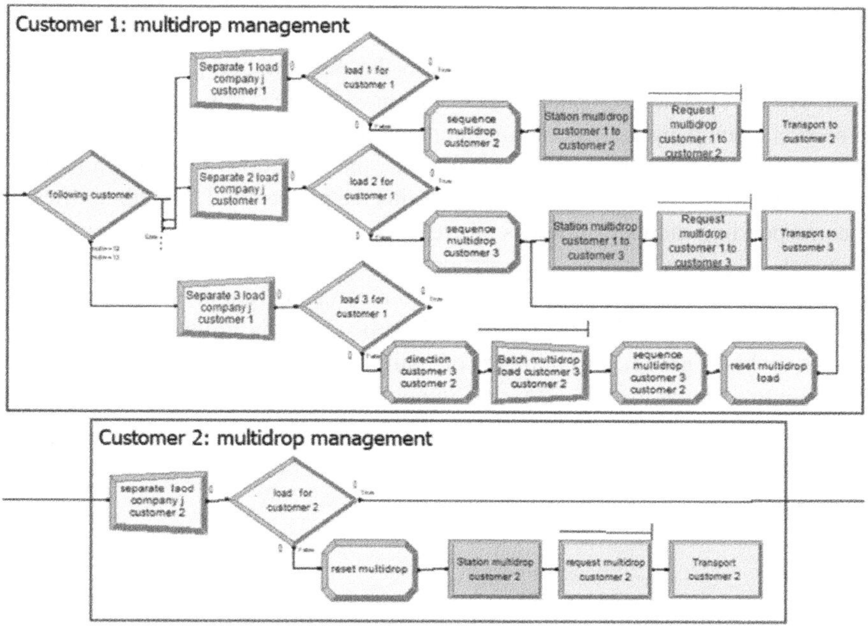

Figure 9.16. *Simulation model of the "Customer i multi-drop management" sections. For a color version of the figure, see www.iste.co.uk/jerbi/discrete.zip*

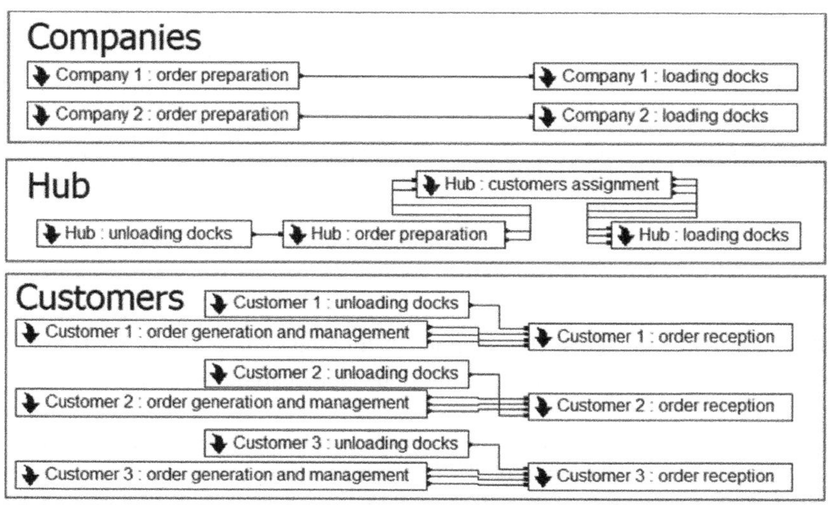

Figure 9.17. *Simulation model of the pooled supply chain with hub*

9.7.4.1. *Companies sub-models*

The companies sub-models are similar to those in the non-pooled supply chain model. However, since companies no longer have a direct link with customers, the sub-section "Company i: Customer assignment" has been removed. Furthermore, in the sub-section "Company i: order preparation", the conditions for holding pallets in stock are modified. Each "Hold" module retains the pallets until the hub notifies that the corresponding number of pallets has reached a minimum level. Moreover, in the "Company i: load preparation" sub-section, there are no longer full or partial loads. Each time the hub sends a product request to a company, the company sends a full load to the hub. Thus, the structure of this sub-section is limited to two "Batch" modules, which group the number of pallets corresponding to the vehicle's capacity into a single load. An "Assign" module then assigns the source of each load (Figure 9.18).

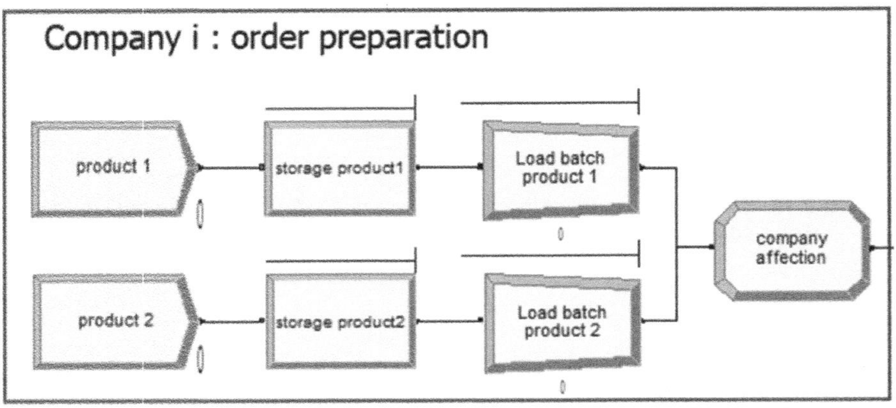

Figure 9.18. *Simulation model of the "company i order preparation" section in the pooled supply chain simulation model with hub. For a color version of the figure, see www.iste.co.uk/jerbi/discrete.zip*

9.7.4.2. *Hub sub-model*

This sub-model includes the sections "Hub: unloading docks", "Hub: order preparation", "Hub: customers assignment" and "Hub: loading docks".

The first two sections are structurally similar to those in the company sub-model of the non-pooled simulation model. The main difference is in the "Hub: order preparation" section. Here, a "Decide" module separates each load into individual pallets to sort them by types. Each type of pallet is stored in the queue of one of the

"Hold" modules, which release them based on signals from customers. Once released, the pallets are sorted according to the load type.

If the load is full, the pallets are sent to the "Batch" module, named "Full load preparation hub", to be grouped and sent to the corresponding customer. Otherwise, the pallets are sent to the "Batch" module named "Partial load preparation hub" to be grouped into a partial load and sent to the corresponding customer (Figure 9.19).

9.7.4.3. *Customer sub-model*

This sub-model is composed of three sections: the "Customer j: order generation and management" section, the "Customer j: unloading docks" section and the "Customer j: order reception" section. The two last sections maintain the same structures as those used in the non-pooled model. The differences are found in the first section. Here, the purchase order is no longer split in two. Instead, each purchase order undergoes a stock verification process at the hub. During this process, several conditions are checked based on the availability of pallet types. The first condition verifies the availability of all pallets. If any required pallet types are missing, the verification extends to the other pallet types of the second company. If all pallet types are unavailable, the order is canceled. Once sorted, the order form passes through various "Signal" modules that send the necessary signals to the "Hold" modules in the "Hub: order preparation" section, prompting the release of the requested pallets (Figure 9.20).

9.8. Verification and validation of simulation models

The simulation models for both the non-pooled and pooled supply chain have been verified. This verification ensures that the model operates according to the conceptual designs by animating the flow of entities during the simulation.

The vaccination center operates as a non-terminating system over a one year period, which is sufficiently long to negate the effects of the transitional period. To ensure robust results, 30 simulation replications are performed for each simulation model. Since the supply chain model is hypothetical, no further validation is required.

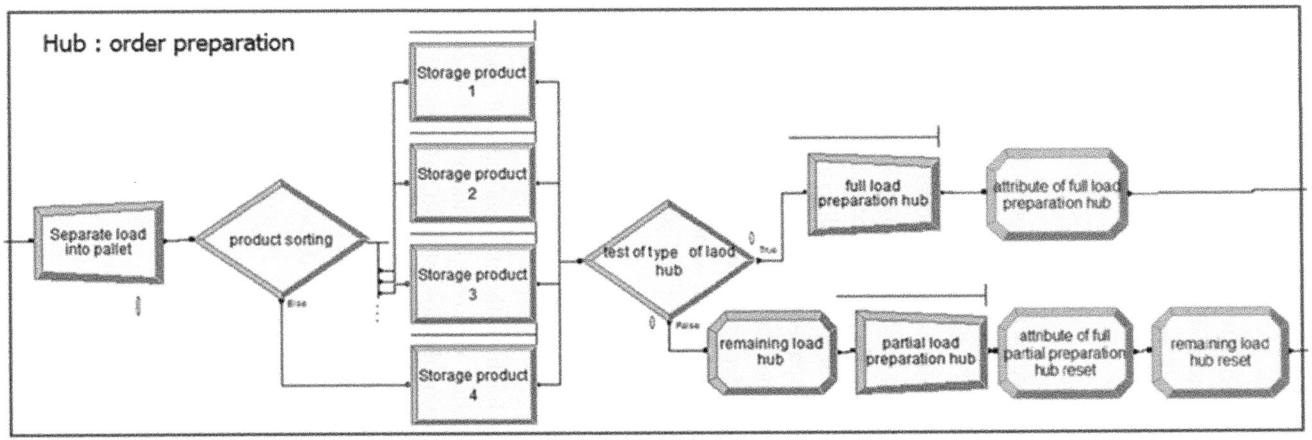

Figure 9.19. *Simulation model of the "Hub order preparation" section in the simulation model of the pooled supply chain with hub*

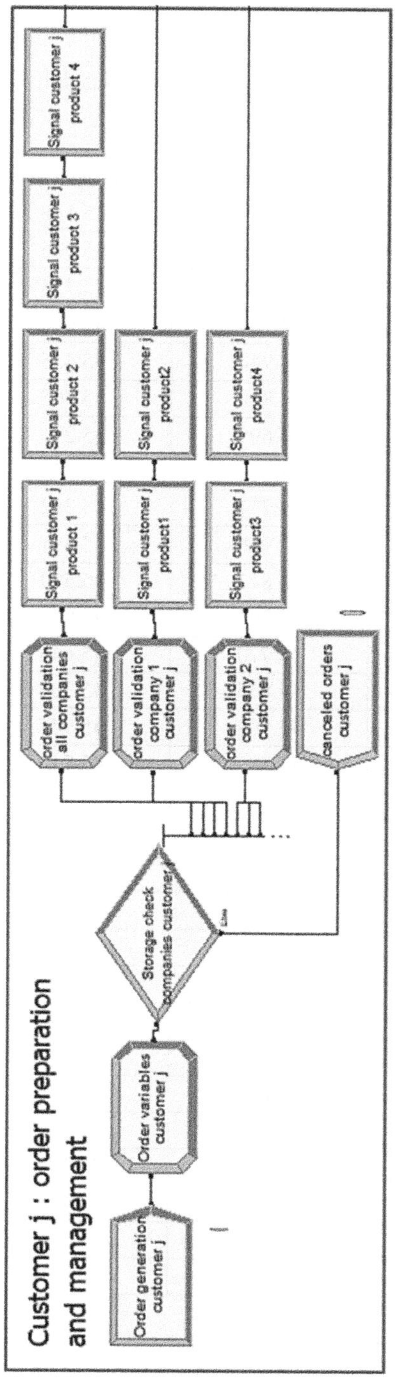

Figure 9.20. *Simulation model of "Customer j order preparation and management" in the supply chain simulation model with hub*

9.9. Results and discussion

Table 9.4 gives the average CO_2 emissions per pallet delivered from the different supply chain models for the 30 replications.

Rep	Non-pooled	Multi-pick	Multi-drop	Hub
1	50.521	44.469	45.186	43.735
2	50.858	44.869	45.536	44.060
3	51.042	44.903	45.723	43.957
4	50.617	44.761	45.620	43.882
5	50.549	44.744	45.435	43.886
6	50.560	44.828	45.505	43.814
7	50.788	44.699	45.270	43.903
8	50.784	44.812	45.853	43.835
9	50.550	44.737	45.282	44.141
10	50.947	44.966	45.491	44.061
11	50.516	44.873	45.579	43.916
12	50.890	44.266	44.924	44.052
13	50.907	44.770	45.646	43.965
14	50.642	44.855	45.360	43.994
15	50.947	45.102	45.948	43.685
16	50.568	44.887	45.248	43.990
17	50.876	44.679	45.857	44.211
18	50.847	44.739	45.470	44.024
19	50.906	44.941	45.943	44.121
20	50.808	44.615	45.266	43.879
21	50.833	44.830	45.575	43.911
22	50.984	44.762	45.271	44.081
23	50.899	44.769	45.422	43.704
24	50.903	44.626	45.076	43.774
25	50.851	45.022	45.558	43.998
26	50.609	44.697	45.361	44.033
27	50.928	44.779	45.790	43.808
28	50.595	44.768	45.428	44.138
29	50.648	44.732	45.215	44.077
30	50.930	44.772	45.448	43.984

Table 9.4. *Simulation results (kgCO_2/pallet)*

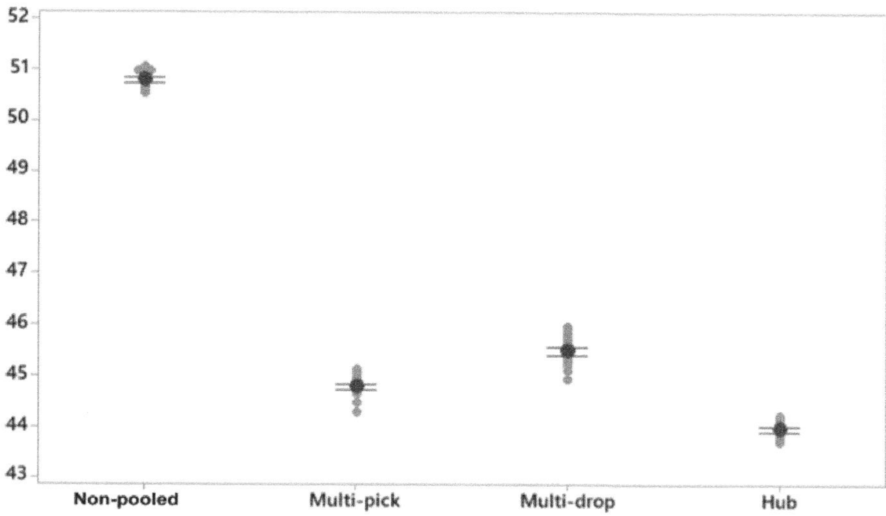

Figure 9.21. *Mean values and confidence intervals of CO_2 emissions*

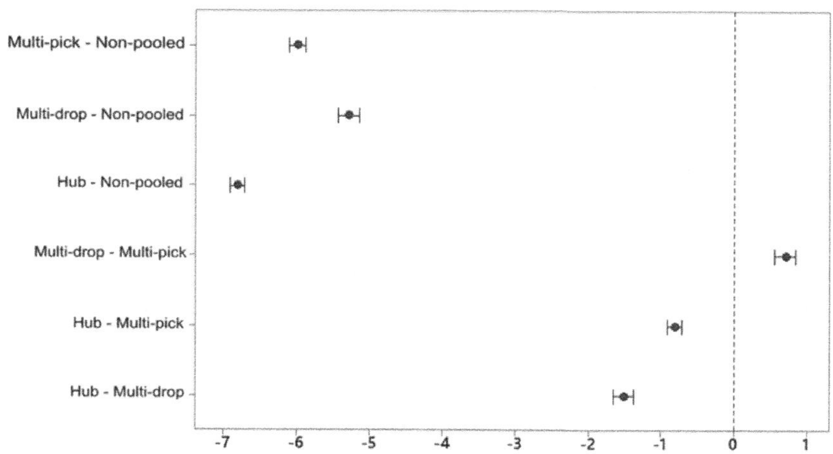

Figure 9.22. *Games–Howell confidence intervals of pairwise differences in CO_2 emissions*

The simulation results were analyzed using Minitab 18 software. Figure 9.21 displays the values and confidence intervals of the average emissions per pallet delivered. It reveals the non-pooled supply chain has the highest average CO_2 emissions, exceeding 50.777 kgCO_2/pallet. In contrast, the various pooling strategies studied have emissions that do not exceed 45.476 $kgCO_2$/pallet. Furthermore, the CO_2 emissions of the different pooling strategies are quite similar.

A pairwise comparison of the simulation results of all strategies was conducted using the Games–Howell method. This procedure is advantageous because it does not assume equal variances among the groups being compared. Figure 9.22 indicates that no confidence interval for the differences between the means of CO_2 emissions for the pooling strategies includes the value zero. Thus, the results of the simulation models are statistically different.

The results of all simulation scenarios are ranked in descending order based on CO_2 emissions in Table 9.5. This summary table reveals that all pooling strategies considerably reduce CO_2 emissions compared to the non-pooled supply chain. The reductions range from 10.439% to 13.437%, with the most substantial reduction observed in the hub strategy.

Strategy	Non-pooled	Hub	Multi-pick	Multi-drop
CO_2 emission ($kgCO_2$/pallet)	50.777	43.954	44.776	45.476
Reduction	Reference	–13.437%	–11.818%	–10.439%

Table 9.5. *Summary of CO_2 emissions results*

9.10. Conclusion

CO_2 emissions stemming from supply chain operations play a critical role in exacerbating the greenhouse effect. Addressing this environmental challenge necessitates effective strategies such as supply chain pooling. This research delves into the impact of various pooling strategies, specifically multi-pick, multi-drop and hub configurations, on CO_2 emissions through advanced simulation techniques.

The study aims to demonstrate how these pooling strategies can mitigate CO_2 emissions within a logistics framework. By analyzing a detailed case study, it became evident that each pooling strategy contributes significantly to reducing CO_2 emissions compared to traditional non-pooled methods. Moreover, the hub pooling strategy emerges as particularly effective, showcasing the highest level of emission reduction among the strategies evaluated.

Overall, this exploration underscores the importance of adopting pooling strategies in supply chains to achieve substantial environmental benefits, with the hub strategy standing out as a particularly promising approach in reducing CO_2 emissions across supply chain operations.

Appendices

Appendix 1

Table of χ^2

ν/P	0.01	0.025	0.05	0.1	0.9	0.95	0.975	0.99
1	0.000	0.001	0.004	0.016	2.706	3.841	5.024	6.635
2	0.020	0.051	0.103	0.211	4.605	5.991	7.378	9.21
3	0.115	0.216	0.352	0.584	6.251	7.815	9.348	11.345
4	0.297	0.484	0.711	1.064	7.779	9.488	11.143	13.277
5	0.554	0.831	1.145	1.610	9.236	11.070	12.833	15.086
6	0.872	1.237	1.635	2.204	10.645	12.592	14.449	16.812
7	1.239	1.690	2.167	2.833	12.017	14.067	16.013	18.475
8	1.646	2.180	2.733	3.490	13.362	15.507	17.535	20.090
9	2.088	2.700	3.325	4.168	14.684	16.919	19.023	21.666
10	2.558	3.247	3.940	4.865	15.987	18.307	20.483	23.209
11	3.053	3.816	4.575	5.578	17.275	19.675	21.920	24.725
12	3.571	4.404	5.226	6.304	18.549	21.026	23.337	26.217
13	4.107	5.009	5.892	7.042	19.812	22.362	24.736	27.688
14	4.660	5.629	6.571	7.790	21.064	23.685	26.119	29.141
15	5.226	6.262	7.261	8.547	22.307	24.996	27.488	30.578

Appendix 2

Table of F for α = 0.05

v2 v1	1	2	3	4	5	6	7	8	9	10	11	12
1	161.45	199.50	215.71	224.58	230.16	233.99	236.77	238.88	240.54	241.88	242.98	243.90
2	18.51	19.00	19.16	19.25	19.30	19.33	19.35	19.37	19.38	19.40	19.40	19.41
3	10.13	9.55	9.28	9.12	9.01	8.94	8.89	8.85	8.81	8.79	8.76	8.74
4	7.71	6.94	6.59	6.39	6.26	6.16	6.09	6.04	6.00	5.96	5.94	5.91
5	6.61	5.79	5.41	5.19	5.05	4.95	4.88	4.82	4.77	4.74	4.70	4.68
6	5.99	5.14	4.76	4.53	4.39	4.28	4.21	4.15	4.10	4.06	4.03	4.00
7	5.59	4.74	4.35	4.12	3.97	3.87	3.79	3.73	3.68	3.64	3.60	3.57
8	5.32	4.46	4.07	3.84	3.69	3.58	3.50	3.44	3.39	3.35	3.31	3.28
9	5.12	4.26	3.86	3.63	3.48	3.37	3.29	3.23	3.18	3.14	3.10	3.07
10	4.96	4.10	3.71	3.48	3.33	3.22	3.14	3.07	3.02	2.98	2.94	2.91
11	4.84	3.98	3.59	3.36	3.20	3.09	3.01	2.95	2.90	2.85	2.82	2.79
12	4.75	3.89	3.49	3.26	3.11	3.00	2.91	2.85	2.80	2.75	2.72	2.69
13	4.67	3.81	3.41	3.18	3.03	2.92	2.83	2.77	2.71	2.67	2.63	2.60
14	4.60	3.74	3.34	3.11	2.96	2.85	2.76	2.70	2.65	2.60	2.57	2.53
15	4.54	3.68	3.29	3.06	2.90	2.79	2.71	2.64	2.59	2.54	2.51	2.48
16	4.49	3.63	3.24	3.01	2.85	2.74	2.66	2.59	2.54	2.49	2.46	2.42
17	4.45	3.59	3.20	2.96	2.81	2.70	2.61	2.55	2.49	2.45	2.41	2.38
18	4.41	3.55	3.16	2.93	2.77	2.66	2.58	2.51	2.46	2.41	2.37	2.34
19	4.38	3.52	3.13	2.90	2.74	2.63	2.54	2.48	2.42	2.38	2.34	2.31
20	4.35	3.49	3.10	2.87	2.71	2.60	2.51	2.45	2.39	2.35	2.31	2.28
21	4.32	3.47	3.07	2.84	2.68	2.57	2.49	2.42	2.37	2.32	2.28	2.25
22	4.30	3.44	3.05	2.82	2.66	2.55	2.46	2.40	2.34	2.30	2.26	2.23
23	4.28	3.42	3.03	2.80	2.64	2.53	2.44	2.37	2.32	2.27	2.24	2.20
24	4.26	3.40	3.01	2.78	2.62	2.51	2.42	2.36	2.30	2.25	2.22	2.18
25	4.24	3.39	2.99	2.76	2.60	2.49	2.40	2.34	2.28	2.24	2.20	2.16
v1: degrees of freedom of the numerator, v2: degrees of freedom of the denominator												

v2 v1	13	14	15	16	17	18	19	20	22	24	26	28	30
1	244.69	245.36	245.95	246.47	246.92	247.32	247.69	248.02	248.58	249.05	249.45	249.80	250.10
2	19.42	19.42	19.43	19.43	19.44	19.44	19.44	19.45	19.45	19.45	19.46	19.46	19.46
3	8.73	8.71	8.70	8.69	8.68	8.67	8.67	8.66	8.65	8.64	8.63	8.62	8.62
4	5.89	5.87	5.86	5.84	5.83	5.82	5.81	5.80	5.79	5.77	5.76	5.75	5.75
5	4.66	4.64	4.62	4.60	4.59	4.58	4.57	4.56	4.54	4.53	4.52	4.50	4.50
6	3.98	3.96	3.94	3.92	3.91	3.90	3.88	3.87	3.86	3.84	3.83	3.82	3.81
7	3.55	3.53	3.51	3.49	3.48	3.47	3.46	3.44	3.43	3.41	3.40	3.39	3.38
8	3.26	3.24	3.22	3.20	3.19	3.17	3.16	3.15	3.13	3.12	3.10	3.09	3.08
9	3.05	3.03	3.01	2.99	2.97	2.96	2.95	2.94	2.92	2.90	2.89	2.87	2.86
10	2.89	2.86	2.85	2.83	2.81	2.80	2.79	2.77	2.75	2.74	2.72	2.71	2.70
11	2.76	2.74	2.72	2.70	2.69	2.67	2.66	2.65	2.63	2.61	2.59	2.58	2.57
12	2.66	2.64	2.62	2.60	2.58	2.57	2.56	2.54	2.52	2.51	2.49	2.48	2.47
13	2.58	2.55	2.53	2.51	2.50	2.48	2.47	2.46	2.44	2.42	2.41	2.39	2.38
14	2.51	2.48	2.46	2.44	2.43	2.41	2.40	2.39	2.37	2.35	2.33	2.32	2.31
15	2.45	2.42	2.40	2.38	2.37	2.35	2.34	2.33	2.31	2.29	2.27	2.26	2.25
16	2.40	2.37	2.35	2.33	2.32	2.30	2.29	2.28	2.25	2.24	2.22	2.21	2.19
17	2.35	2.33	2.31	2.29	2.27	2.26	2.24	2.23	2.21	2.19	2.17	2.16	2.15
18	2.31	2.29	2.27	2.25	2.23	2.22	2.20	2.19	2.17	2.15	2.13	2.12	2.11
19	2.28	2.26	2.23	2.21	2.20	2.18	2.17	2.16	2.13	2.11	2.10	2.08	2.07
20	2.25	2.22	2.20	2.18	2.17	2.15	2.14	2.12	2.10	2.08	2.07	2.05	2.04
21	2.22	2.20	2.18	2.16	2.14	2.12	2.11	2.10	2.07	2.05	2.04	2.02	2.01
22	2.20	2.17	2.15	2.13	2.11	2.10	2.08	2.07	2.05	2.03	2.01	2.00	1.98
23	2.18	2.15	2.13	2.11	2.09	2.08	2.06	2.05	2.02	2.01	1.99	1.97	1.96
24	2.15	2.13	2.11	2.09	2.07	2.05	2.04	2.03	2.00	1.98	1.97	1.95	1.94
25	2.14	2.11	2.09	2.07	2.05	2.04	2.02	2.01	1.98	1.96	1.95	1.93	1.92
v1: degrees of freedom of the numerator, v2: degrees of freedom of the denominator													

Appendix 3

Duncan Multiplier Table for α = 0.05

ddl\f	2	3	4	5	6	7	8	9	10
24	2.92	3.07	3.16	3.23	3.28	3.31	3.35	3.37	3.39
30	2.89	3.03	3.13	3.20	3.25	3.29	3.32	3.35	3.37
40	2.86	3.01	3.10	3.17	3.22	3.27	3.30	3.33	3.35
60	2.83	2.98	3.07	3.14	3.20	3.24	3.28	3.31	3.33
120	2.80	2.95	3.04	3.12	3.17	3.22	3.25	3.29	3.31
∞	2.77	2.92	3.02	3.09	3.15	3.19	3.23	3.27	3.29

Appendix 4

UTIL (in %)

Rep	Actual model				Proposed model			
	Registration	Control	Vaccination	Validation	Registration	Control	Vaccination	Validation
1	22.21	51.77	52.56	44.17	50.73	39.06	66.83	22.21
2	21.01	50.67	49.70	41.92	51.16	36.86	62.85	21.01
3	21.24	49.45	51.09	41.42	49.73	37.56	62.20	21.24
4	22.13	51.56	49.24	43.53	53.27	39.36	66.20	22.13
5	21.61	50.93	50.42	42.48	51.28	37.81	64.42	21.61
6	22.54	52.15	50.56	43.65	50.97	39.13	65.67	22.54
7	21.93	53.95	51.80	44.76	53.91	39.85	67.91	21.93
8	22.72	52.04	49.50	44.02	51.30	38.95	67.02	22.72
9	21.81	49.36	47.88	41.60	50.45	37.58	62.28	21.81
10	20.86	50.42	48.02	41.43	49.91	36.13	63.11	20.86
11	22.48	52.26	50.93	43.79	51.51	38.64	66.32	22.48
12	21.78	52.54	50.35	43.82	51.93	37.60	66.40	21.78
13	22.31	51.99	51.05	43.25	51.42	38.57	65.09	22.31
14	21.92	52.24	48.55	44.07	52.27	38.49	65.04	21.92
15	23.36	53.29	53.37	45.53	53.94	39.78	68.09	23.36
16	21.51	51.19	50.19	43.22	50.58	37.58	64.54	21.51
17	20.56	49.16	48.89	41.40	49.88	37.19	63.55	20.56
18	22.16	50.73	48.97	43.24	51.41	38.36	64.78	22.16
19	21.52	52.33	50.28	43.13	51.39	37.86	65.33	21.52
20	21.81	50.97	49.62	43.08	51.86	37.52	63.68	21.81
21	22.44	52.98	49.91	43.87	51.37	37.90	63.40	22.44
22	21.36	51.12	50.89	42.46	51.03	37.53	63.11	21.36
23	21.79	51.18	49.47	42.16	50.73	36.12	61.63	21.79
24	21.33	51.55	49.67	41.79	50.03	37.40	64.18	21.33
25	20.87	59.71	48.72	41.05	49.01	36.48	62.02	20.87
26	21.91	61.39	50.14	42.23	51.09	38.45	64.14	21.91
27	21.68	62.59	50.19	42.60	50.84	37.44	65.17	21.68
28	20.58	61.02	48.61	41.02	49.37	35.60	63.12	20.58
29	20.89	59.97	47.67	41.20	49.66	36.18	61.60	20.89
30	21.31	61.00	48.02	42.31	49.73	36.68	61.85	21.31

Table A4.1. *Theoretical arrivals*

Rep	Actual model				Proposed model			
	Registration	Control	Vaccination	Validation	Registration	Control	Vaccination	Validation
1	21.87	52.11	52.21	43.81	32.30	52.43	38.91	64.98
2	22.38	53.02	49.60	43.55	32.45	51.58	37.57	66.90
3	20.96	49.17	50.15	43.54	32.75	51.17	36.03	62.97
4	22.13	51.56	49.24	43.53	33.21	53.27	39.36	66.20
5	22.44	51.84	51.44	44.35	33.63	51.33	38.35	65.41
6	21.62	52.06	51.97	44.85	33.19	51.42	38.93	66.05
7	21.54	49.28	50.09	42.29	31.97	50.78	37.56	64.61
8	21.09	48.64	47.63	40.96	31.70	48.13	36.85	63.20
9	21.34	52.61	52.60	43.93	33.25	53.00	39.40	65.38
10	20.87	47.46	47.12	41.02	31.39	49.69	37.00	61.79
11	22.33	49.83	50.15	42.69	32.35	50.55	37.93	64.27
12	21.43	49.87	50.10	43.08	31.74	51.43	36.83	64.14
13	21.78	49.86	50.70	42.65	33.13	50.15	36.83	62.42
14	22.59	51.17	49.71	43.87	33.00	52.29	38.88	64.91
15	21.25	50.57	48.58	40.99	31.83	49.35	36.72	62.05
16	21.98	51.65	50.59	43.15	32.17	50.77	37.90	63.90
17	22.09	52.29	52.32	44.49	33.16	51.39	39.81	65.91
18	22.12	50.47	50.22	42.96	33.79	50.65	37.50	64.47
19	21.62	50.68	50.54	42.58	31.90	51.09	38.17	64.33
20	21.43	48.99	47.99	40.67	31.39	48.88	37.49	62.74
21	20.94	51.56	51.70	43.00	31.94	50.43	37.82	64.43
22	22.80	55.83	53.50	45.65	33.60	54.80	40.45	68.29
23	21.48	47.80	48.52	40.89	31.31	50.06	36.55	63.28
24	21.84	51.78	49.45	42.53	32.85	50.11	37.36	62.94
25	21.51	54.21	50.95	42.97	31.89	50.61	38.85	64.63
26	22.85	54.47	49.53	43.28	34.17	52.74	38.83	65.72
27	22.04	53.74	50.03	43.59	32.66	51.49	37.77	66.66
28	21.83	53.17	50.74	42.81	31.88	52.20	38.80	65.08
29	21.51	53.46	50.38	41.90	32.08	50.33	36.72	62.87
30	22.63	53.66	52.51	43.29	33.61	51.85	38.01	65.96

Table A4.2. *Hypothetical arrivals*

Rep	Real model				Proposed model			
	Registration	Control	Vaccination	Validation	Registration	Control	Vaccination	Validation
1	21.54	51.25	50.53	42.15	32.16	51.35	38.56	63.40
2	21.65	50.82	49.51	42.94	32.52	51.18	36.93	62.64
3	20.81	49.17	49.69	42.52	30.93	49.10	35.92	61.80
4	21.71	50.09	51.31	43.49	33.03	51.57	38.72	64.71
5	21.06	50.43	49.27	41.98	32.57	51.64	37.50	64.02
6	22.63	53.14	53.19	44.29	33.80	53.22	38.48	67.63
7	21.36	49.93	49.50	41.75	31.15	49.83	37.24	62.67
8	22.11	52.63	51.32	43.43	33.00	52.94	38.87	66.64
9	21.46	49.01	49.00	40.55	29.99	48.60	35.48	61.73
10	21.31	51.24	49.87	42.63	31.75	50.79	37.03	63.96
11	20.30	49.19	48.22	40.94	30.56	46.43	34.87	62.31
12	21.91	52.28	50.71	42.52	32.45	50.30	36.83	64.20
13	21.13	50.92	47.83	42.03	32.12	50.67	37.03	62.67
14	21.09	49.66	47.75	41.75	32.49	49.90	37.97	63.73
15	21.19	51.93	48.26	43.41	32.44	52.27	38.20	65.14
16	21.19	47.81	48.26	43.41	32.44	52.27	38.20	65.14
17	19.53	51.93	45.76	40.45	30.06	47.53	35.37	59.43
18	20.72	49.24	47.74	41.62	31.62	49.93	36.29	62.14
19	20.41	49.08	46.87	41.61	31.40	49.47	36.60	60.32
20	22.37	51.71	52.36	44.37	33.22	53.95	39.27	65.20
21	21.46	49.73	49.91	43.15	32.52	50.03	37.85	63.68
22	21.75	49.80	47.89	41.56	31.53	49.10	37.81	63.55
23	22.61	52.18	50.78	44.07	33.10	51.97	38.56	66.33
24	21.73	51.58	50.66	43.51	32.38	50.89	38.89	64.15
25	20.44	51.65	47.17	40.00	30.34	47.44	35.59	59.96
26	22.25	54.21	52.78	45.11	33.84	53.98	38.04	67.48
27	22.10	54.07	51.19	44.39	34.35	53.41	40.01	67.84
28	21.14	52.19	47.30	41.03	30.32	48.77	36.75	62.29
29	20.65	52.33	48.55	41.09	30.83	48.69	35.85	60.97
30	20.73	53.27	49.35	40.91	32.32	49.67	37.17	62.98

Table A4.3. *Actual arrivals*

References

Addi Ait, H. (2000). *Optimisation des flux de production : méthodes et simulation*. Dunod, Paris.

Altiok, T. and Melamed, B. (2007). *Simulation and Analysis with Arena*. Elsevier, Amsterdam.

Arena (2018). Arena template reference guide. Rockwell Software Inc.

Chung, C.A. (2003). *Simulation Modeling Handbook: A Practical Approach*. CRC Press, Boca Raton.

Fleury, G., Lacomme, P., Tanguy, A. (2006). *Simulation à évènements discrets avec CD-Rom. Modèles déterministes et stochastique*. Eyrolles, Paris.

Foucart, T. (2000). *Introduction aux tests statistiques*. Technip, Paris.

Kelton, D., Sadowski, R., Zupick, N. (2015). *Simulation with Arena*. McGraw-Hill Education, New York.

Polenghi, A., Fumagalli, L., Roda, I. (2018). Role of simulation in industrial engineering: Focus on manufacturing systems. *IFAC-PapersOnLine*, 51(11), 496–501.

Ycart, A. (2002). *Estimation paramétrique tests statistiques*. Centre de publication universitaire, Tunis.

Index

A, B

C, D

E, F

H, M

O, P

R, S

T, V

Other titles from

iSTE

in

Computer Engineering

2024

CHAKRABORTY Rajdeep, GHOSH Anupam, KUMAR MANDAL Jyotsna, CHOUDHURY Tanupriya, CHATTERJEE Prasenjit
Smart Edge Computing: An Operation Research Perspective
(International Perspectives in Decision Analytics and Operations Research Set – Volume 2)

HOMÈS Bernard
Fundamentals of Software Testing (2nd edition, revised and updated)

PRADHAN Devasis, KUMAR SAHU Prasanna, MYO TUN Hla, CHATTERJEE Prasenjit
Artificial and Cognitive Computing for Sustainable Healthcare Systems in Smart Cities
(International Perspectives in Decision Analytics and Operations Research Set – Volume 3)

2023

CHOUDHURY Tanupriya, KHANNA Abhirup, CHATTERJEE Prasenjit, UM Jung-Sup, BHATTACHARYA Abhishek
Blockchain Applications in Healthcare: Innovations and Practices (International Perspectives in Decision Analytics and Operations Research Set – Volume 1)

PAI G A Vijayalakshmi
A Textbook of Data Structures and Algorithms 1: Mastering Linear Data Structures
A Textbook of Data Structures and Algorithms 2: Mastering Nonlinear Data Structures
A Textbook of Data Structures and Algorithms 3: Mastering Advanced Data Structures and Algorithm Design Strategies

2022

HOMÈS Bernard
Advanced Testing of Systems-of-Systems 1: Theoretical Aspects
Advanced Testing of Systems-of-Systems 2: Practical Aspects

MEHTA Shikha, TIWARI Sanju, SIARRY Patrick, JABBAR M.A.
Tools, Languages, Methodologies for Representing Semantics on the Web of Things

SIDHOM Sahbi, KADDOUR Amira
Systems and Uses of Digital Sciences for Knowledge Organization (Digital Tools and Uses Set – Volume 9)

ZAIDOUN Ameur Salem
Computer Science Security: Concepts and Tools

2021

DELHAYE Jean-Loïc
Inside the World of Computing: Technologies, Uses, Challenges

HARDIN Thérèse, JAUME Mathieu, PESSAUX François, VIGUIÉ DONZEAU-GOUGE Véronique
Concepts and Semantics of Programming Languages 1: A Semantical Approach with OCaml and Python
Concepts and Semantics of Programming Languages 2: Modular and Object-oriented Constructs with OCaml, Python, C++, Ada and Java

MKADMI Abderrazak
Archives in The Digital Age: Preservation and the Right to be Forgotten (Digital Tools and Uses Set – Volume 8)

TOKLU Yusuf Cengiz, BEKDAŞ Gebrail, NIGDELI Sinan Melih
Metaheuristics for Structural Design and Analysis (Optimization Heuristics Set – Volume 3)

2020

DARCHE Philippe
Microprocessor 1: Prolegomenes – Calculation and Storage Functions – Models of Computation and Computer Architecture
Microprocessor 2: Core Concepts – Communication in a Digital System
Microprocessor 3: Core Concepts – Hardware Aspects
Microprocessor 4: Core Concepts – Software Aspects
Microprocessor 5: Software and Hardware Aspects of Development, Debugging and Testing – The Microcomputer

DUVAUT Patrick, DALLOZ Xavier, MENGA David, KOEHL François, CHRIQUI Vidal, BRILL Joerg
Internet of Augmented Me, I.AM: Empowering Innovation for a New Sustainable Future

LAFFLY Dominique
TORUS 1 – Toward an Open Resource Using Services: Cloud Computing for Environmental Data
TORUS 2 – Toward an Open Resource Using Services: Cloud Computing for Environmental Data
TORUS 3 – Toward an Open Resource Using Services: Cloud Computing for Environmental Data

LAURENT Anne, LAURENT Dominique, MADERA Cédrine
Data Lakes
(Databases and Big Data Set – Volume 2)

OULHADJ Hamouche, DAACHI Boubaker, MENASRI Riad
Metaheuristics for Robotics
(Optimization Heuristics Set – Volume 2)

SADIQUI Ali
Computer Network Security

VENTRE Daniel
Artificial Intelligence, Cybersecurity and Cyber Defense

2019

BESBES Walid, DHOUIB Diala, WASSAN Niaz, MARREKCHI Emna
Solving Transport Problems: Towards Green Logistics

CLERC Maurice
Iterative Optimizers: Difficulty Measures and Benchmarks

GHLALA Riadh
Analytic SQL in SQL Server 2014/2016

TOUNSI Wiem
Cyber-Vigilance and Digital Trust: Cyber Security in the Era of Cloud Computing and IoT

2018

ANDRO Mathieu
Digital Libraries and Crowdsourcing
(Digital Tools and Uses Set – Volume 5)

ARNALDI Bruno, GUITTON Pascal, MOREAU Guillaume
Virtual Reality and Augmented Reality: Myths and Realities

CARDON Alain
Beyond Artificial Intelligence: From Human Consciousness to Artificial Consciousness

HOMAYOUNI S. Mahdi, FONTES Dalila B.M.M.
Metaheuristics for Maritime Operations
(Optimization Heuristics Set – Volume 1)

JEANSOULIN Robert
JavaScript and Open Data

PIVERT Olivier
NoSQL Data Models: Trends and Challenges
(Databases and Big Data Set – Volume 1)

SEDKAOUI Soraya
Data Analytics and Big Data

SALEH Imad, AMMI Mehdi, SZONIECKY Samuel
Challenges of the Internet of Things: Technology, Use, Ethics
(Digital Tools and Uses Set – Volume 7)

SZONIECKY Samuel
Ecosystems Knowledge: Modeling and Analysis Method for Information and Communication
(Digital Tools and Uses Set – Volume 6)

2017

BENMAMMAR Badr
Concurrent, Real-Time and Distributed Programming in Java: Threads, RTSJ and RMI

BOUHAÏ Nasreddine, SALEH Imad
Internet of Things: Evolutions and Innovations
(Digital Tools and Uses Set – Volume 4)

HÉLIODORE Frédéric, NAKIB Amir, ISMAIL Boussaad, OUCHRAA Salma, SCHMITT Laurent
Metaheuristics for Intelligent Electrical Networks
(Metaheuristics Set – Volume 10)

MA Haiping, SIMON Dan
Evolutionary Computation with Biogeography-based Optimization
(Metaheuristics Set – Volume 8)

PÉTROWSKI Alain, BEN-HAMIDA Sana
Evolutionary Algorithms
(Metaheuristics Set – Volume 9)

PAI G A Vijayalakshmi
Metaheuristics for Portfolio Optimization: An Introduction using MATLAB®
(Metaheuristics Set – Volume 11)

REYES-GARCIA Everardo
The Image-Interface: Graphical Supports for Visual Information
(Digital Tools and Uses Set – Volume 3)

REYES-GARCIA Everardo, BOUHAÏ Nasreddine
Designing Interactive Hypermedia Systems
(Digital Tools and Uses Set – Volume 2)

SZONIECKY Samuel, BOUHAÏ Nasreddine
Collective Intelligence and Digital Archives: Towards Knowledge Ecosystems
(Digital Tools and Uses Set – Volume 1)

2016

BLUM Christian, FESTA Paola
Metaheuristics for String Problems in Bio-informatics
(Metaheuristics Set – Volume 6)

DEROUSSI Laurent
Metaheuristics for Logistics
(Metaheuristics Set – Volume 4)

DHAENENS Clarisse, JOURDAN Laetitia
Metaheuristics for Big Data
(Metaheuristics Set – Volume 5)

LABADIE Nacima, PRINS Christian, PRODHON Caroline
Metaheuristics for Vehicle Routing Problems
(Metaheuristics Set – Volume 3)

LEROY Laure
Eyestrain Reduction in Stereoscopy

LUTTON Evelyne, PERROT Nathalie, TONDA Alberto
Evolutionary Algorithms for Food Science and Technology
(Metaheuristics Set – Volume 7)

MAGOULÈS Frédéric, ZHAO Hai-Xiang
Data Mining and Machine Learning in Building Energy Analysis

RIGO Michel
Advanced Graph Theory and Combinatorics

ZELM Martin, DOUMEINGTS Guy, MENDONÇA Joao Pedro
Enterprise Interoperability: in the Digitized and Networked Factory of the Future

2015

BARBIER Franck, RECOUSSINE Jean-Luc
COBOL Software Modernization: From Principles to Implementation with the BLU AGE® Method

CHEN Ken
Performance Evaluation by Simulation and Analysis with Applications to Computer Networks

CLERC Maurice
Guided Randomness in Optimization
(Metaheuristics Set – Volume 1)

DURAND Nicolas, GIANAZZA David, GOTTELAND Jean-Baptiste, ALLIOT Jean-Marc
Metaheuristics for Air Traffic Management
(Metaheuristics Set – Volume 2)

MAGOULÈS Frédéric, ROUX François-Xavier, HOUZEAUX Guillaume
Parallel Scientific Computing

MUNEESAWANG Paisarn, YAMMEN Suchart
Visual Inspection Technology in the Hard Disk Drive Industry

2014

BOULANGER Jean-Louis
Formal Methods Applied to Industrial Complex Systems
Formal Methods Applied to Complex Systems: Implementation of the B Method

EXPOSITO Ernesto, DIOP Codé
Smart SOA Platforms in Cloud Computing Architectures

GARDI Frédéric, BENOIST Thierry, DARLAY Julien, ESTELLON Bertrand, MEGEL Romain
Mathematical Programming Solver based on Local Search

KRICHEN Saoussen, CHAOUACHI Jouhaina
Graph-related Optimization and Decision Support Systems

LARRIEU Nicolas, VARET Antoine
Rapid Prototyping of Software for Avionics Systems: Model-oriented Approaches for Complex Systems Certification

LAURAS Matthieu, ZELM Martin, ARCHIMÈDE Bernard, BÉNABEN Frédérick, DOUMEINGTS Guy
Enterprise Interoperability: Interoperability for Agility, Resilience and Plasticity of Collaborations

OUSSALAH Mourad Chabane
Software Architecture 1
Software Architecture 2

QUESNEL Flavien
Scheduling of Large-scale Virtualized Infrastructures: Toward Cooperative Management

RIGO Michel
Formal Languages, Automata and Numeration Systems 1: Introduction to Combinatorics on Words
Formal Languages, Automata and Numeration Systems 2: Applications to Recognizability and Decidability

SAINT-DIZIER Patrick
Musical Rhetoric: Foundations and Annotation Schemes

TOUATI Sid, DUPONT DE DINECHIN Benoit
Advanced Backend Code Optimization

2013

ANDRÉ Etienne, SOULAT Romain
The Inverse Method: Parametric Verification of Real-time Embedded Systems

BOULANGER Jean-Louis
Safety Management for Software-based Equipment

DELAHAYE Daniel, PUECHMOREL Stéphane
Modeling and Optimization of Air Traffic

FRANCOPOULO Gil
LMF — Lexical Markup Framework

GHEDIRA Khaled
Constraint Satisfaction Problems: CSP Formalisms and Techniques

GOVAERT Gérard, NADIF Mohamed
Co-Clustering: Models, Algorithms and Applications

LANDRAGIN Frédérick
Man–Machine Dialogue: Design and Challenges

LUCAS Laurent, LOSCOS Céline, REMION Yannick
3D Video: Models, From Capture to Diffusion

ROCHANGE Christine, UHRIG Sascha, SAINRAT Pascal
Time-Predictable Architectures

WAHBI Mohamed
Algorithms and Ordering Heuristics for Distributed Constraint Satisfaction Problems

ZELM Martin, VAN SINDEREN Marten, FERRAIRA PIRES Luis, DOUMEINGTS Guy
Enterprise Interoperability: Research and Applications in the Service-oriented Ecosystem

2012

ARBOLEDA Hugo, ROYER Jean-Claude
Model-Driven and Software Product Line Engineering

BLANCHET Gérard, DUPOUY Bertrand
Computer Architecture

BOULANGER Jean-Louis
Industrial Use of Formal Methods: Formal Verification
Formal Methods: Industrial Use from Model to the Code

CALVARY Gaëlle, DELOT Thierry, SÈDES Florence, TIGLI Jean-Yves
Computer Science and Ambient Intelligence

MAHOUT Vincent
Assembly Language Programming: ARM Cortex-M3 2.0: Organization, Innovation and Territory

MARLET Renaud
Program Specialization

SOTO Maria, SEVAUX Marc, ROSSI André, LAURENT Johann
Memory Allocation Problems in Embedded Systems: Optimization Methods

2011

BICHOT Charles-Edmond, SIARRY Patrick
Graph Partitioning

BOULANGER Jean-Louis
Static Analysis of Software: The Abstract Interpretation

CAFERRA Ricardo
Logic for Computer Science and Artificial Intelligence

KORDON Fabrice, HADDAD Serge, PAUTET Laurent, PETRUCCI Laure
Distributed Systems: Design and Algorithms
Models and Analysis in Distributed Systems

LORCA Xavier
Tree-based Graph Partitioning Constraint

RIDHA MAHJOUB A.
Progress in Combinatorial Optimization: ISCO2010

TRUCHET Charlotte, ASSAYAG Gerard
Constraint Programming in Music

VICAT-BLANC Pascale, GOGLIN Brice, GUILLIER Romaric,
SOUDAN Sébastien
Computing Networks: From Cluster to Cloud Computing

2010

AUBRUN Christophe, SIMON Daniel, SONG Ye-Qiong
Co-design Approaches for Dependable Networked Control Systems

AUDIBERT Pierre
Mathematics for Informatics and Computer Science

BABAU Jean-Philippe, BLAY-FORNARINO Mireille, CHAMPEAU Joël,
ROBERT Sylvain, SABETTA Antonino
Model Driven Engineering for Distributed Real-Time Systems: MARTE modeling, model transformations and their usages

BOULANGER Jean-Louis
Safety of Computer Architectures

SIGAUD Olivier, BUFFET Olivier
Markov Decision Processes in Artificial Intelligence

MONMARCHÉ Nicolas, GUINAND Frédérick, SIARRY Patrick
Artificial Ants

SOLNON Christine
Ant Colony Optimization and Constraint Programming

2009

FOURNIER Jean-Claude
Graph Theory and Applications: with exercises and problems

GUÉDON Jeanpierre
The Mojette Transform: theory and applications

JARD Claude, ROUX Olivier H.
Communicating Embedded Systems: Software and Design

LECOUTRE Christophe
Constraint Networks: Techniques and Algorithms

2008

BANÂTRE Michel, MARRON Pedro Jose, OLLERO Anibal, WOLITZ Adam
Cooperating Embedded Systems and Wireless Sensor Networks

MERZ Stephan, NAVET Nicolas
Modeling and Verification of Real-time Systems: Formalisms and Software Tools

PASCHOS Vangelis Th
Combinatorial Optimization and Theoretical Computer Science: Interfaces and Perspectives

WALDNER Jean-Baptiste
Nanocomputers and Swarm Intelligence

2007

BENHAMOU Frédéric, JUSSIEN Narendra, O'SULLIVAN Barry
Trends in Constraint Programming

JUSSIEN Narendra
A to Z of Sudoku

2006

BABAU Jean-Philippe, CHAMPEAU Joël, GÉRARD Sébastien
From MDD Concepts to Experiments and Illustrations – DRES 2006

HABRIAS Henri, FRAPPIER Marc
Software Specification Methods: An Overview Using a Case Study

MURAT Cécile, PASCHOS Vangelis Th
Probabilistic Combinatorial Optimization on Graphs

PANETTO Hervé, BOUDJLIDA Nacer
Interoperability for Enterprise Software and Applications: Proceedings of the Workshops and the Doctorial Symposium of the Second IFAC/IFIP I-ESA International Conference: EI2N, WSI, IS-TSPQ 2006

2005

GÉRARD Sébastien, BABAU Jean-Philippe, CHAMPEAU Joël
Model Driven Engineering for Distributed Real Time Embedded Systems

PANETTO Hervé
Interoperability of Enterprise Software and Applications: Workshops of the INTEROP-ESA 2005 International Conference (EI2N, WSI, ISIDI and IEHENA 2005)

Printed and bound by CPI Group (UK) Ltd, Croydon, CR0 4YY
13/09/2026
14961049-0004